Learning and Applying Solid Edge V20 Step-by-Step

L. Scott Hansen, Ph.D.

Associate Professor of Engineering Technology
College of Computing, Integrated Engineering and Technology
Southern Utah University
Cedar City, Utah

Industrial Press, Inc
New York

ISBN-13 (978-0-8311-)3312-2

First Edition, November 2007

Sponsoring Editor: John Carleo
Cover Design: Janet Romano

Industrial Press Inc.
989 Avenue of the Americas, 19th Floor
New York, NY 10018

10 9 8 7 6 5 4 3 2 1

Note to the Reader

This book provides clear and concise applied instruction in order to help you develop a mastery of *Solid Edge*. Almost every instruction includes a graphic illustration to aid in clarifying that instruction. Software commands appear in **bold** or in "quotation marks" for anyone who prefers not to read every word of the text. Most illustrations also include small pointer arrows and text to further clarify instructions.

This book was written for classroom instruction or self-study, including for individuals with no solid modeling experience at all. You will begin at a very basic level, but by the time you finish you will be completing complex functions.

For any organization requiring additional help, I am available for onsite training. Please contact me at hansens@suu.edu

Scott Hansen
Cedar City, Utah

Table of Contents

Chapter 1 Getting Started

Objectives:

- Create a simple sketch using the Sketch commands
- Dimension a sketch using the SmartDimension command
- Extrude a sketch using the Protrusion command
- Create a hole using the Cutout command
- Create a fillet using the Round command
- Create a counter bore using the Hole command

Chapter 1 includes instruction on how to design the part shown below.

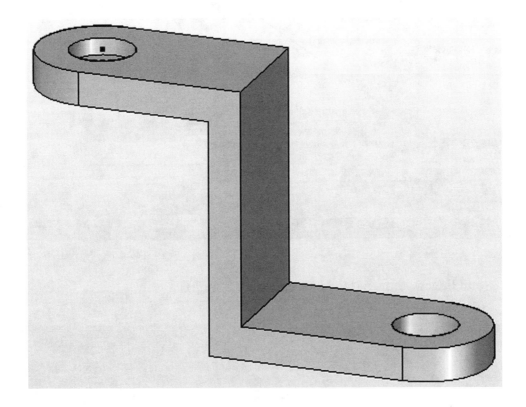

1. Start Solid Edge by moving the cursor to the 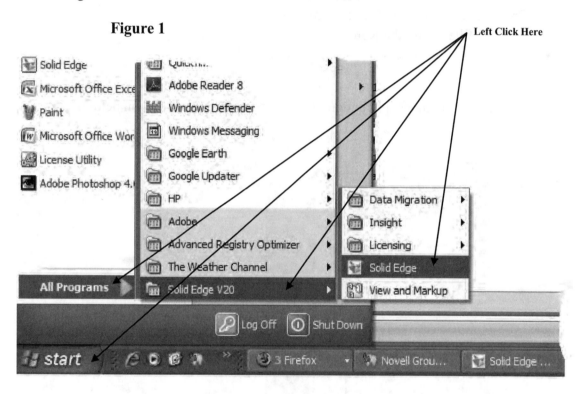 **start** button in the lower left corner of the screen. Click the left mouse button once.

2. A pop up menu of the programs that are installed on the computer will appear. Scroll through the list of programs until you find "Solid Edge".

3. Move the cursor over the text "Solid Edge" and left click once as shown in Figure 1.

Figure 1

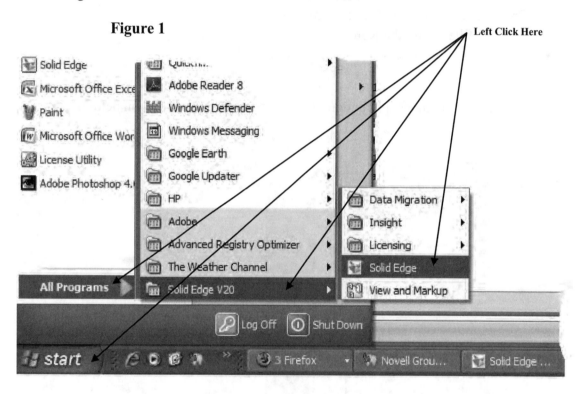

4. Solid Edge will open (load up and begin running).

5. The Text on Buttons warning dialog box may appear indicating that if the screen resolution is lower than 1280 x 1024, toolbars containing text on buttons may exceed the display area. If this is a problem you will need to increase the screen resolution to 1280 x 1024. If this is not a problem left click on **Yes** as shown in Figure 2.

Figure 2

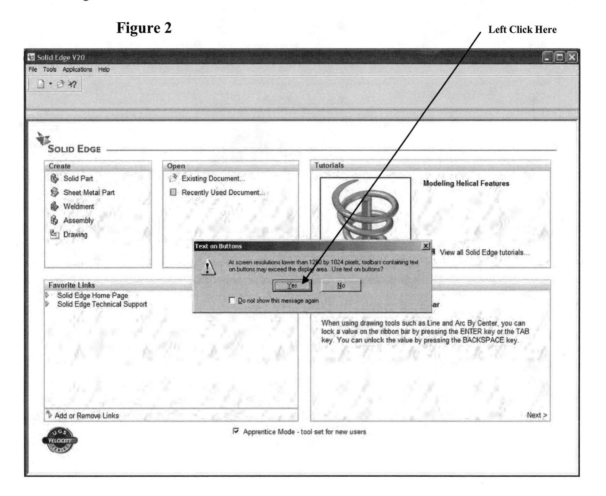

6.   The screen should look similar to Figure 3.

   Figure 3

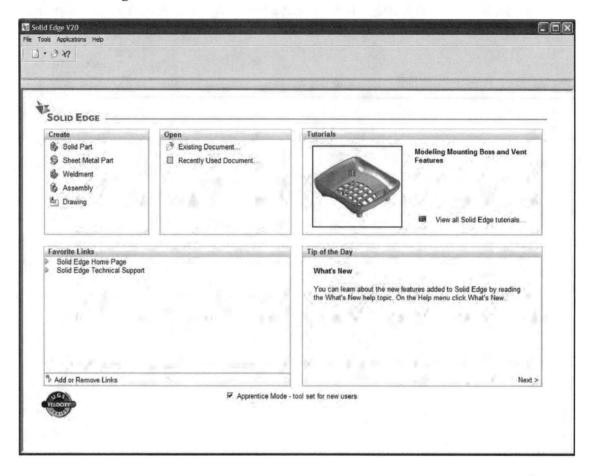

7.   Move the cursor to the upper left portion of the screen and left click on the
      Solid Part link as shown in Figure 4.

Figure 4

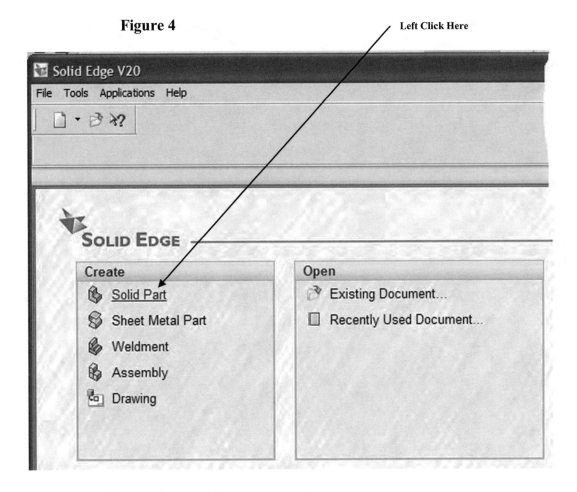

8.   Move the cursor to the upper middle portion of the screen and left click on the "Sketch" icon as shown in Figure 5.

Figure 5

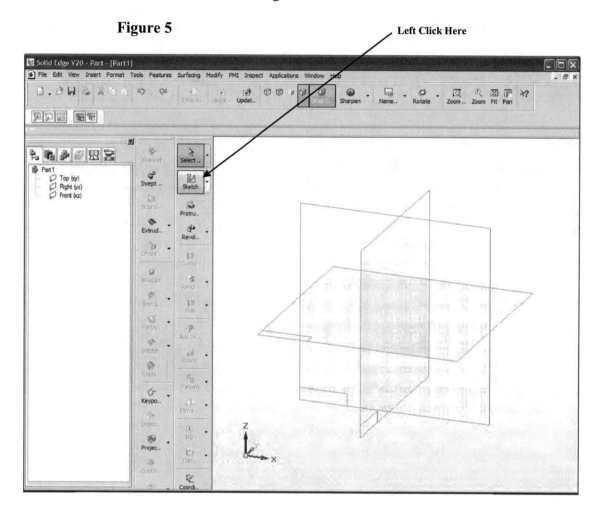

9.　Solid Edge will display three different work planes. Move the cursor over the Front or "xz" work plane until it turns red and left click as shown in Figure 6.

Figure 6

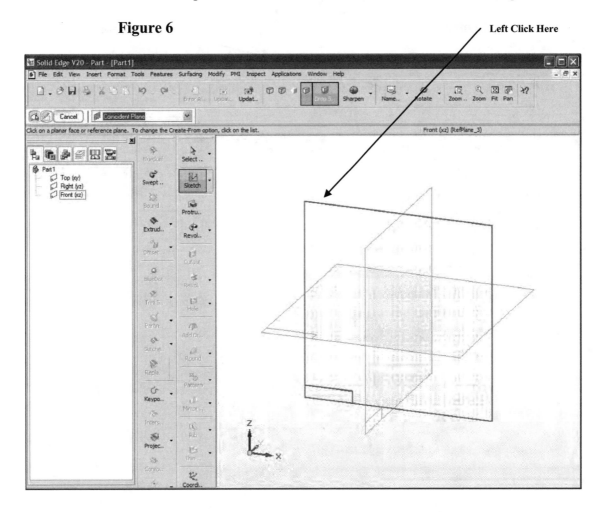

10.　The screen should look similar to Figure 7. If the gird is not visible, left click on Grid as shown.

Figure 7

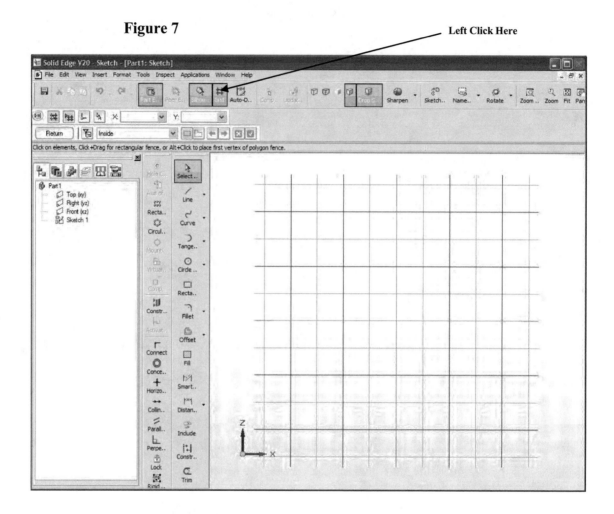

11.　Solid Edge is now ready for use. Move the cursor to the upper middle portion of the screen and left click on **Line** as shown in Figure 8.

Figure 8

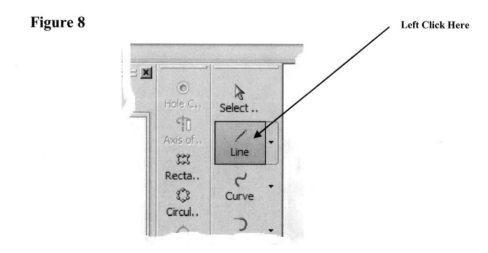

12.   Move the cursor in the lower left portion of the screen and left click once. This will be the beginning end point of a line as shown in Figure 9.

Figure 9

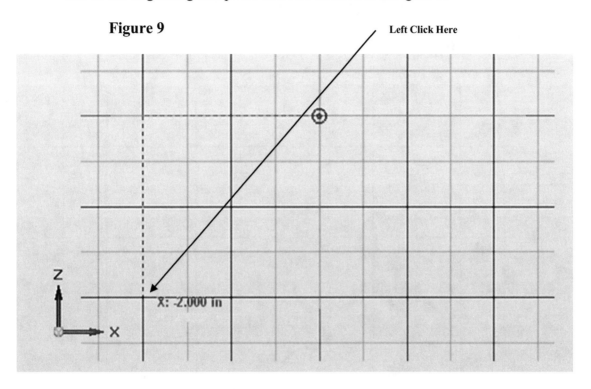

13.   Move the cursor towards the lower right portion of the screen and left click once as shown in Figure 10.

Figure 10

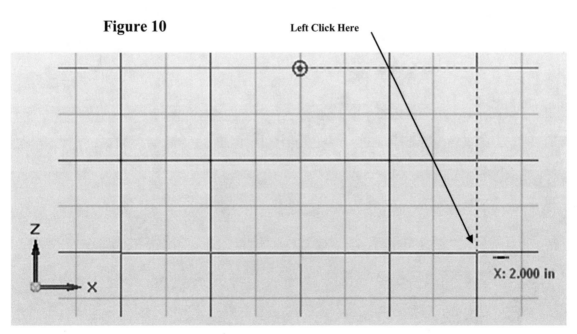

14. While the line is still attached to the cursor, move the cursor towards the top of the screen and left click once. Notice the length of the line is attached to the cursor at the right as shown in Figure 11.

Figure 11

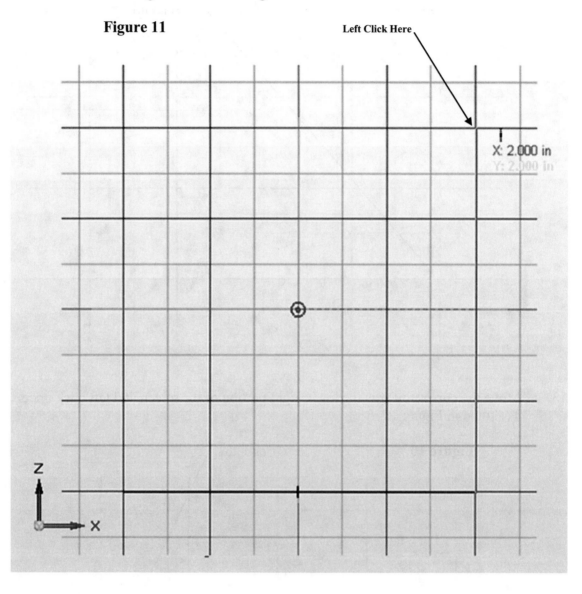

15. With the line still attached to the cursor, move the cursor towards the left side of the screen. Notice the line of small dashes connecting the first and last point together. Left click once when the small dashes appear as shown in Figure 12.

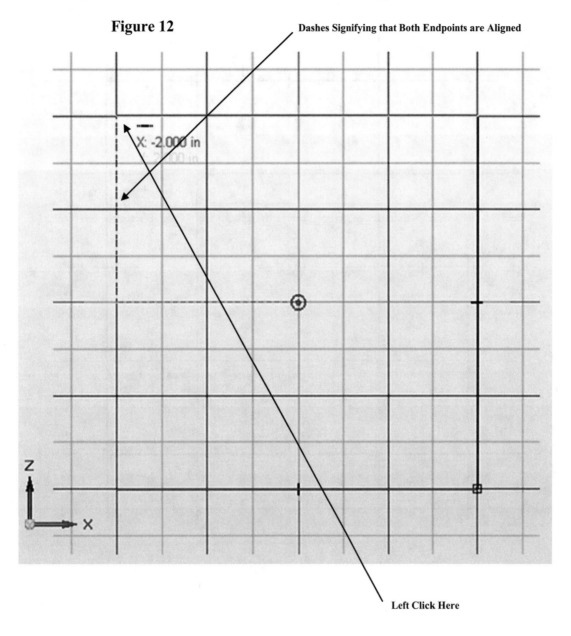

Figure 12

Dashes Signifying that Both Endpoints are Aligned

X: -2.000 in

Z

X

Left Click Here

16.   The screen should look similar to Figure 13.

Figure 13

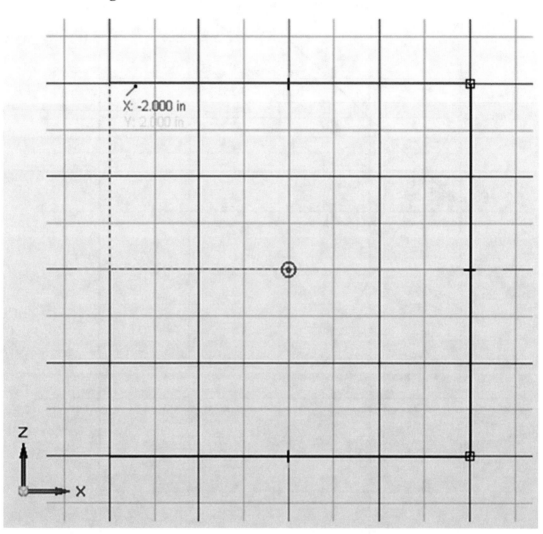

17. Move the cursor down towards the original starting point and left click once. This will form a 90 degree box as shown in Figure 14. **For future reference, pressing the right mouse button will disconnect the endpoint of the line from the cursor.**

Figure 14

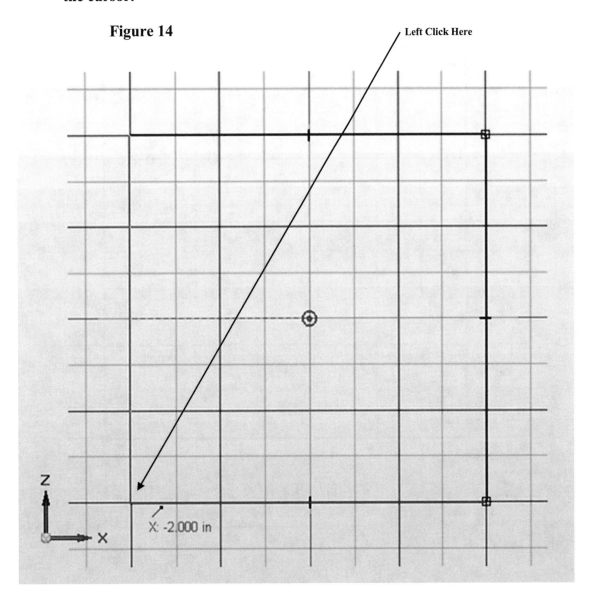

18. The screen should look similar to Figure 15.

Figure 15

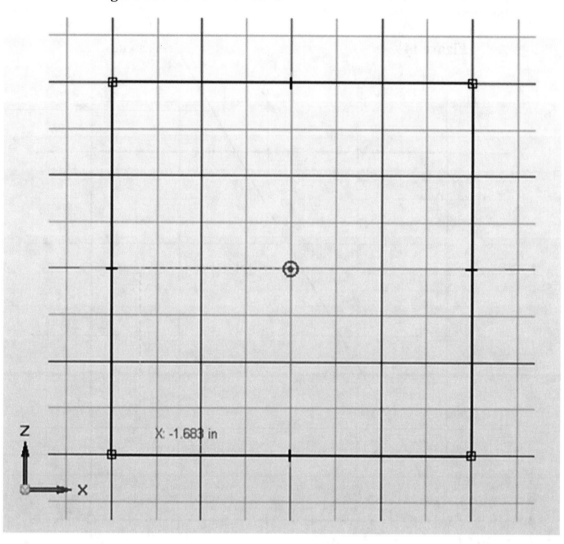

19. To change the dimension style or units, move the cursor to the upper left portion of the screen and left click on **Format**. A drop down menu will appear. Left click on **Style** as shown in Figure 16.

Figure 16

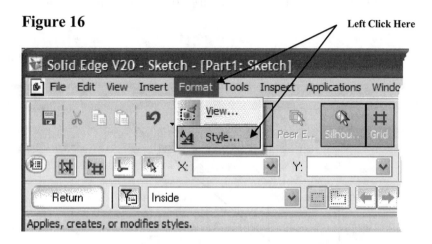

20. The **Style** dialog box will appear. Left click on **Modify** as shown in Figure 17.

Figure 17   Left Click Here

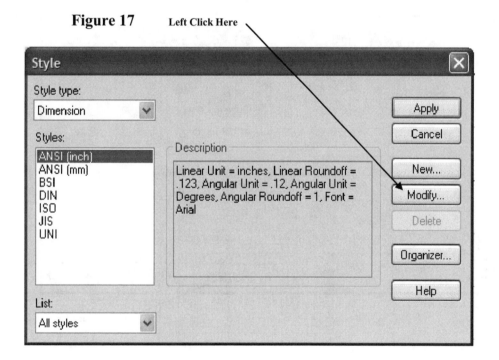

15

21.   The Modify Dimension Style dialog box will appear. Left click on the **Units** tab. Left click on the drop down arrow to the right of Units. If the units are not set to inches then select inches from the drop down menu and left click on **OK** as shown in Figure 18. If the units are already set to inches then left click on **Cancel**.

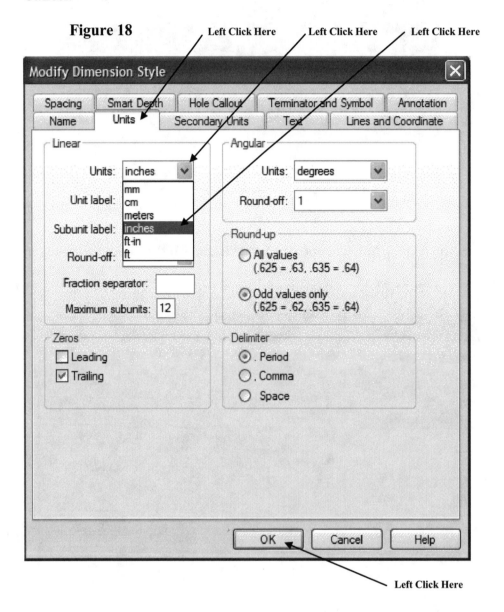

Figure 18

Left Click Here   Left Click Here   Left Click Here

Left Click Here

22. Move the cursor to the upper left portion of the screen and left click on **SmartDimension** as shown in Figure 19.

Figure 19

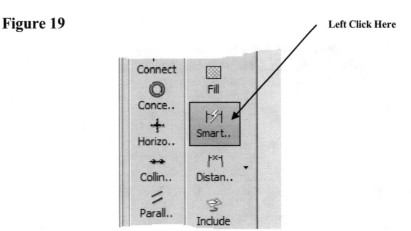

23. Move the cursor over the bottom horizontal line until it turns red as shown in Figure 20. Select the line by left clicking anywhere on the line. Left click once. The dimension will be attached to the cursor.

Figure 20

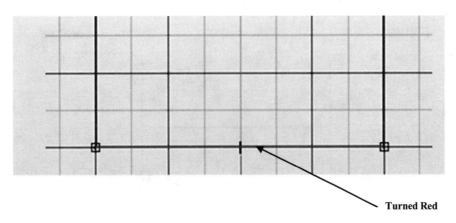

24. Move the cursor down to where the dimension will be placed and left click once as shown in Figure 21.

Figure 21

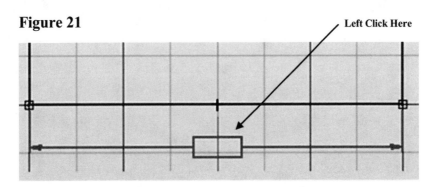

25. The dimension will appear at the upper left portion of the screen as shown in Figure 22.

Figure 22

Enter 2.000 Here

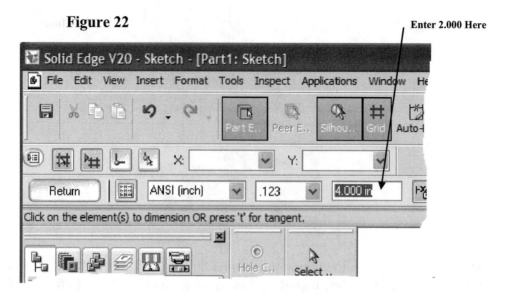

26. While the dimension is highlighted type **2.000** in the dimension box and press the **Enter** key on the keyboard. The dimension of the line will become 2.000 inches as shown in Figure 23.

Figure 23

Length is 2.000 Inches

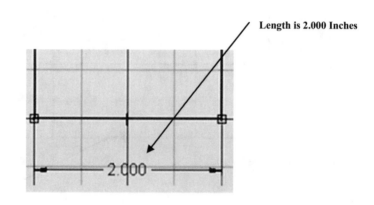

27.	Move the cursor over the right side vertical line until it turns red as shown in Figure 24. Left click once.

Figure 24

Turned Red

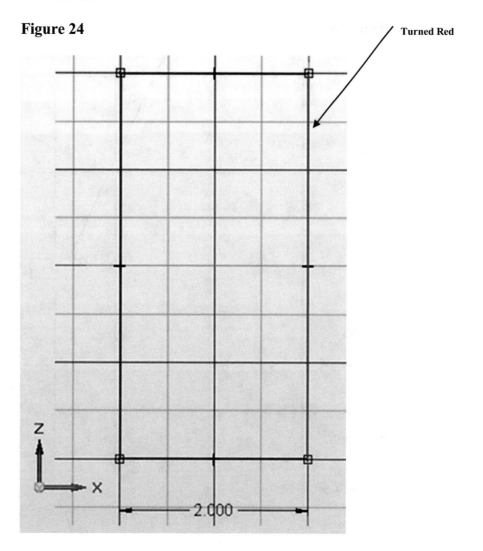

28. Move the cursor to where the dimension will be placed and left click once as shown in Figure 25.

Figure 25

Left Click Here

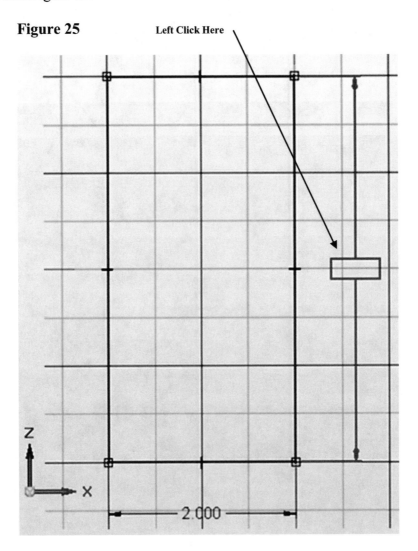

29. The dimension will appear at the upper left portion of the screen as shown in Figure 26.

Figure 26

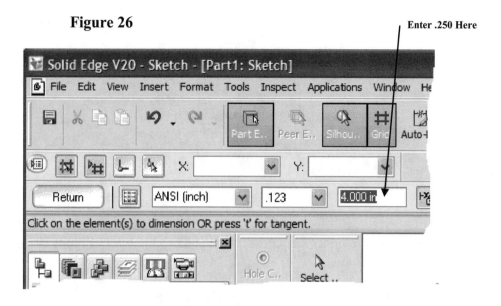

Enter .250 Here

30. While the dimension is highlighted type **.250** in the dimension box and press the **Enter** key on the keyboard. The dimension of the line will become .250 inches as shown in Figure 27.

Figure 27

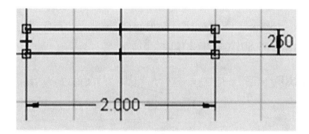

31. Move the cursor to the upper left portion of the screen and left click on **Line** as shown in Figure 28.

Figure 28   Left Click Here

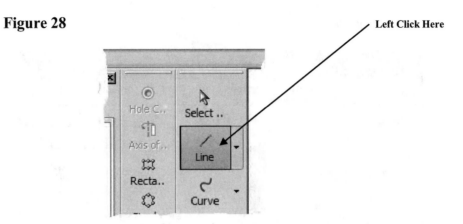

32. Move the cursor to the upper left corner of the box as shown in Figure 29 and left click once.

Figure 29   Left Click Here

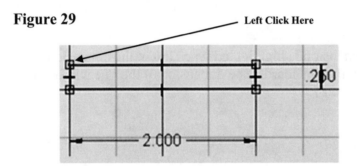

33. Move the cursor upward to create a vertical line and left click once. Right click to disconnect the line as shown in Figure 30.

Figure 30   Left Click Here Then Right Click

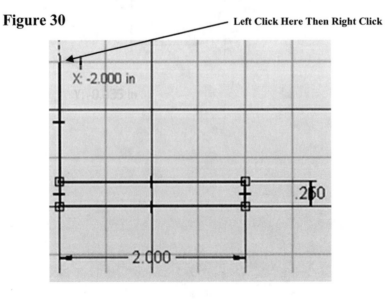

34. You may have to zoom out to leave enough room to construct the following lines. To zoom in or out use the scroll wheel on the mouse or move the cursor to the upper right portion of the screen and left click on the **Zoom** icon as shown in Figure 31.

Figure 31

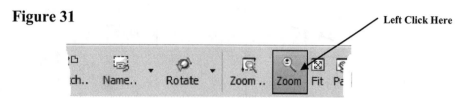

35. After selecting the Zoom icon, the cursor will change to a magnifying glass. Move the cursor to the middle of the screen. Hold the left mouse button down and drag the cursor up to zoom out or down to zoom in.

36. Move the cursor to the upper left corner of the screen and left click on **Line** as shown in Figure 32.

Figure 32

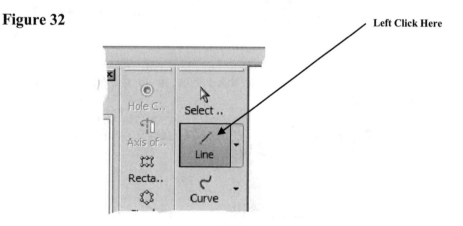

37. Left click on the upper endpoint of the line. Move the cursor to the left and left click once as shown in Figure 33.

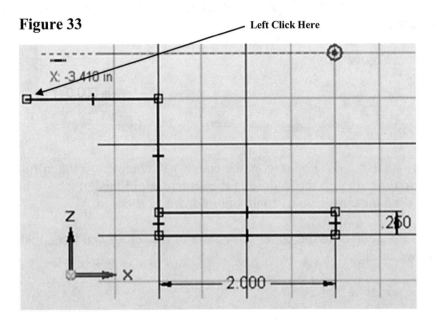

Figure 33

Left Click Here

38. With the line still attached to the cursor, move the cursor up and left click once as shown in Figure 34.

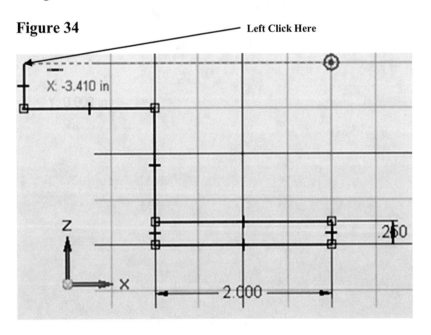

Figure 34

Left Click Here

39. With the line still attached to the cursor, move the cursor to the right side of the screen and left click once as shown in Figure 35.

Figure 35

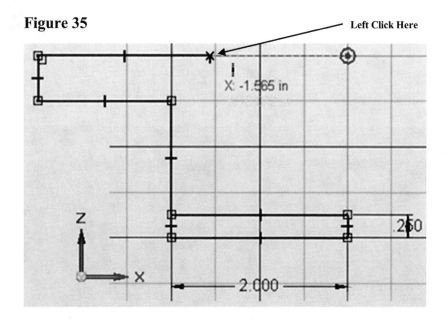

40. With the line still attached to the cursor, move the cursor down towards the bottom of the screen and left click once. After the line has snapped to the horizontal line right click once as shown in Figure 36.

Figure 36

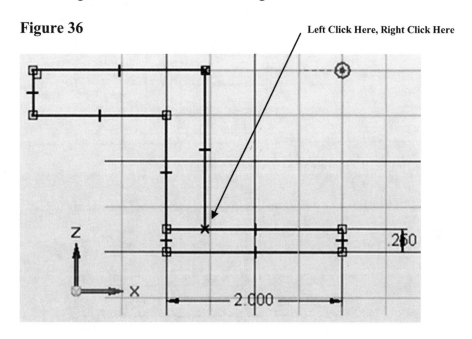

41. Move the cursor to the lower left portion of the screen and left click on the **Trim** command as shown in Figure 37.

Figure 37

Left Click Here

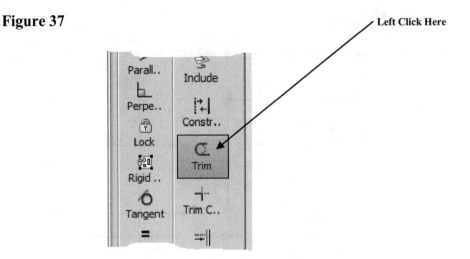

42. Move the cursor over the line that will be trimmed causing it to turn red and left click once as shown in Figure 38.

Figure 38

Left Click Here

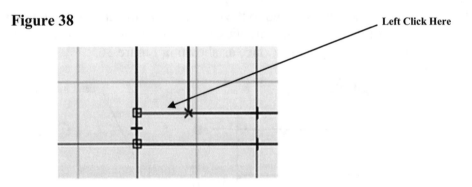

43. The line will disappear as shown in Figure 39.

Figure 39

Line is Trimmed

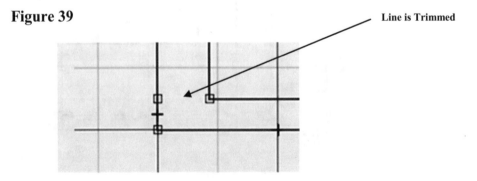

44.   Move the cursor over the line in the lower left corner of the drawing as shown in Figure 40. The line will turn red. This particular line must be deleted so that the line above can be extended the full length.

Figure 40   Line To Be Deleted

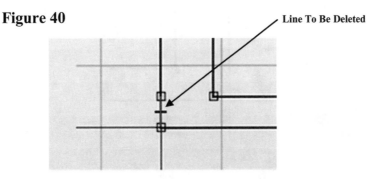

45.   Left click on the line causing it to turn yellow. Move the cursor to the upper left portion of the screen and left click on **Edit**. Left click on **Delete** as shown in Figure 41.

Figure 41   Left Click Here

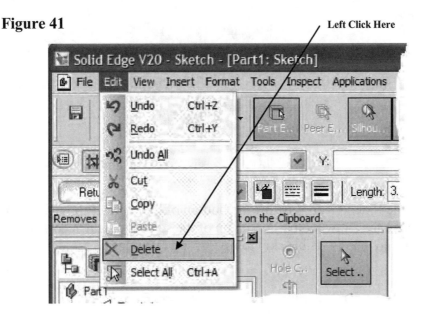

46.   The line will be deleted as shown in Figure 42.

Figure 42

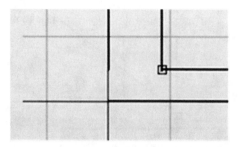

47.   Move the cursor to the lower middle portion of the screen and left click on **Extend** as shown in Figure 43.

Figure 43   **Left Click Here**

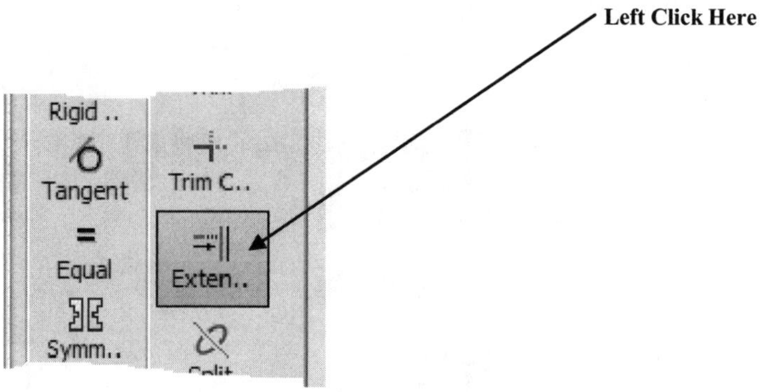

48.   If the Extend icon is below the bottom of the screen (not visible) you will need to move the Draw toolbar to the top of the screen to gain access to the icon. If the Extend icon is visible skip to instruction number 55.

49. To move the Draw toolbar to the top of the screen left click (holding the left mouse button down) on the line above the Select icon as shown in Figure 44.

Figure 44

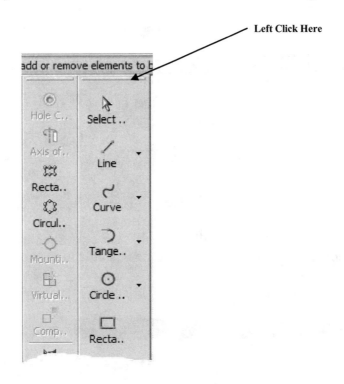

50. Holding the left mouse button down, move the cursor to the location below the "Update" icon and release the left mouse button as shown in Figure 45.

Figure 45

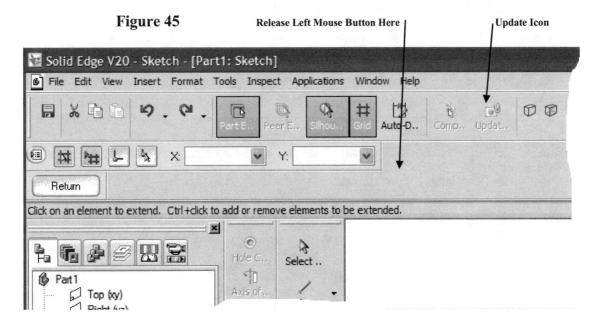

51. The screen should look similar to Figure 46. Notice the Draw toolbar is now located below the "Update" icon.

Figure 46   Draw Toolbar in New Location

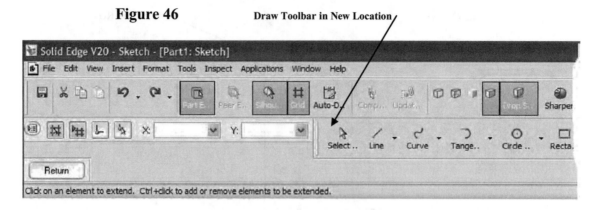

52. If the Extend icon is still not visible left click (holding the left mouse button down) on the line to the left of the Select icon as shown in Figure 47.

Figure 47   Left Click Here

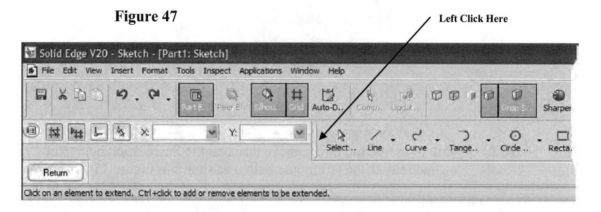

53. Holding the left mouse button down, move the cursor below the "Return " icon and release the left mouse button as shown in Figure 48.

Figure 48   Return Icon   Draw Toolbar in New Location   Extend Icon Now Visible

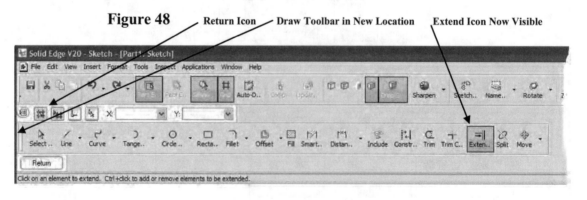

54. Notice the Extend icon is now visible as shown in Figure 48.

55. Move the cursor over the line to be extended (the lower portion). It will turn red and provide a preview of the extend as shown in Figure 49.

Figure 49

Turned Red/Preview

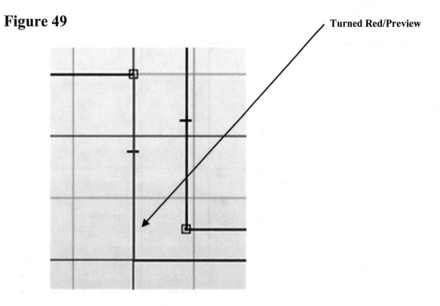

56. After the line has turned red, left click once. The line will extend to intersect the other line as shown in Figure 50.

Figure 50

Left Click Here

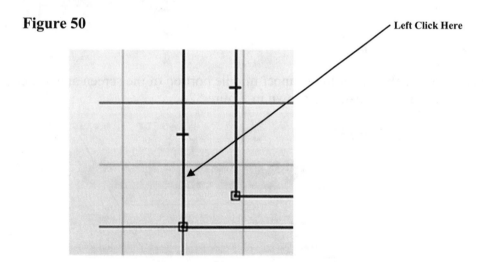

57. The screen should look similar to Figure 51.

Figure 51

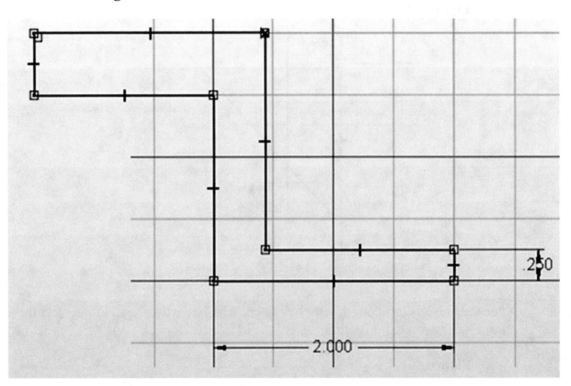

58. Move the cursor to the upper middle portion of the screen and left click on **SmartDimension** as shown in Figure 52.

Figure 52

Left Click Here

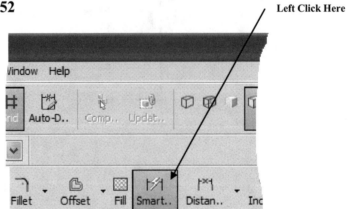

59.   After selecting **SmartDimension** move the cursor over the left side vertical line. The line will turn red as shown in Figure 53. Left click on the line.

Figure 53

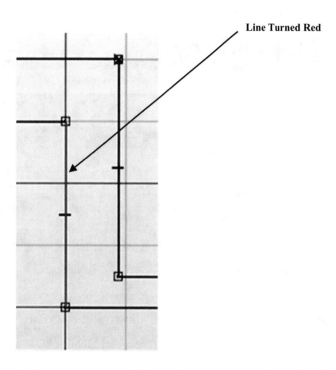

Line Turned Red

60.   The dimension is attached to the cursor. Move the cursor to where the dimension will be placed and left click once.

Figure 54

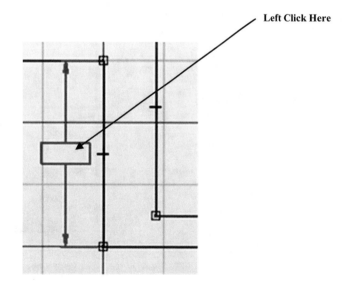

Left Click Here

61. While the dimension is highlighted type **1.750** in the dimension box as shown in Figure 55. Press the **Enter** key on the keyboard.

Figure 55

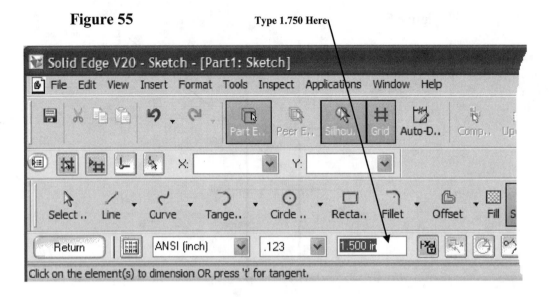

62. The dimension is now 1.750 as shown in Figure 56.

Figure 56

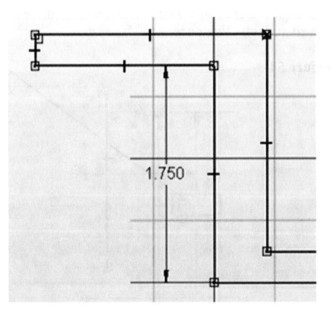

63. After selecting **SmartDimension** move the cursor over the left side vertical line. The line will turn red as shown in Figure 57. Left click on the line.

Figure 57

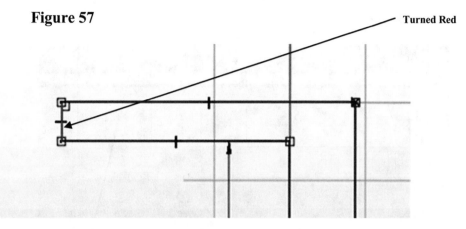

Turned Red

64. The dimension is attached to the cursor. Move the cursor to where the dimension will be placed and left click once as shown in Figure 58.

Figure 58

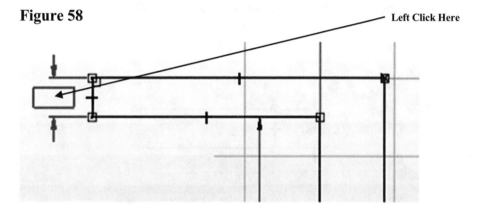

Left Click Here

65. While the dimension is highlighted type **.250** in the dimension box as shown in Figure 59. Press the **Enter** key on the keyboard.

Figure 59

Type .250 Here

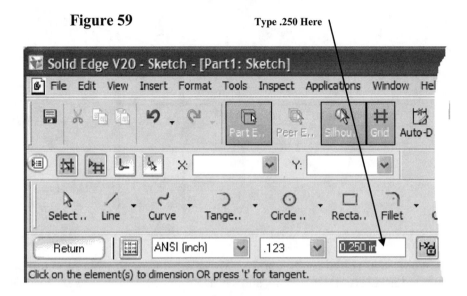

66. The dimension is now .250 as shown in Figure 60.

Figure 60

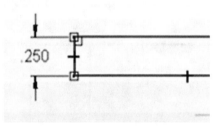

67. Move the cursor over the top horizontal line until it turns red as shown in Figure 61. Left click on the line.

Figure 61

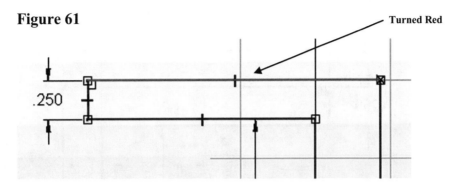

68. The dimension is attached to the cursor. Move the cursor to where the dimension will be placed and left click once as shown in Figure 62.

Figure 62

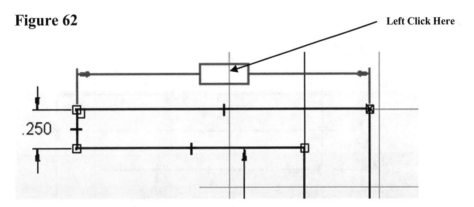

69. While the dimension is highlighted type **1.750** in the dimension box as shown in Figure 63. Press the **Enter** key on the keyboard.

Figure 63

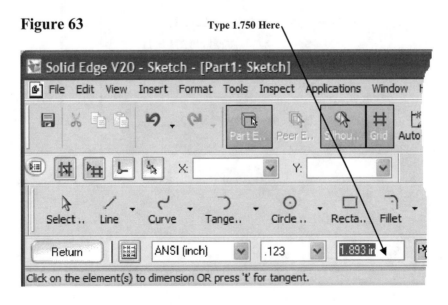

70.   The dimension is now 1.750 as shown in Figure 64.

Figure 64

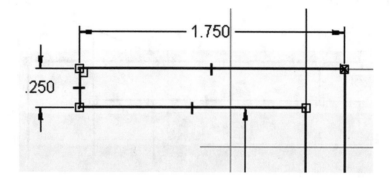

71.   Move the cursor over to the left side vertical line until it turns red and left click as shown in Figure 65.

Figure 65

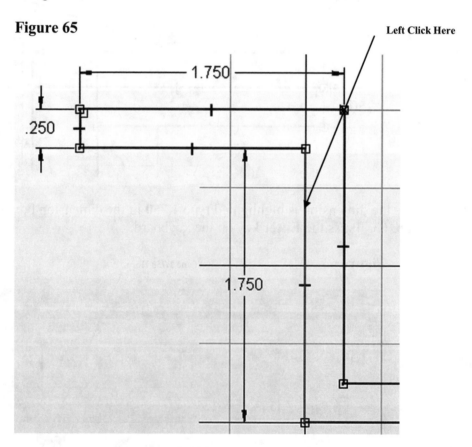

Left Click Here

72. Move the cursor to the other vertical line until it turns red and left click as shown in Figure 66.

Figure 66

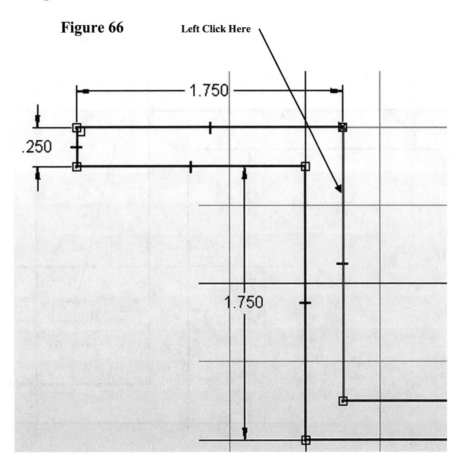

73. The dimension is attached to the cursor. Move the cursor to where the dimension will be placed and left click once as shown in Figure 67.

Figure 67

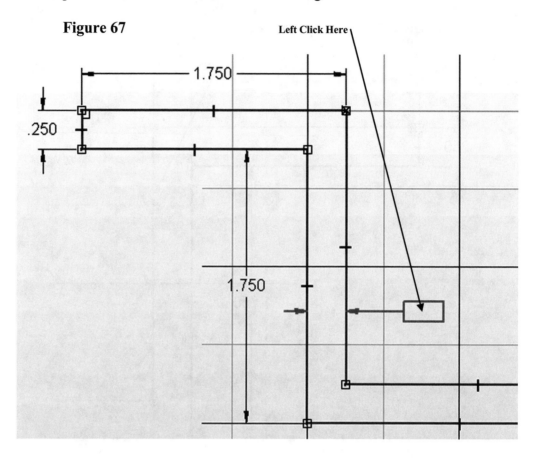

74. While the dimension is highlighted type **.250** in the dimension box as shown in Figure 68. Press the **Enter** key on the keyboard. The dimension of the line will become .250 inches as shown in Figure 69. Press the **Esc** key several times to exit the SmartDimension command.

Figure 68

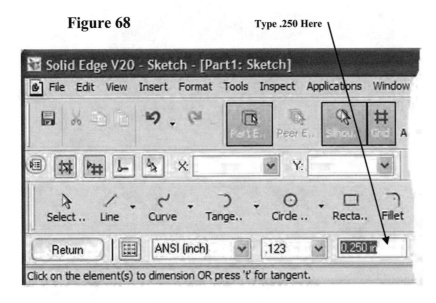

75. The screen should look similar to Figure 69.

Figure 69

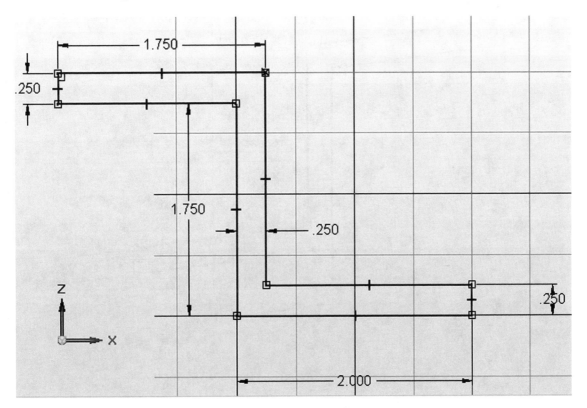

76. Move the cursor to the upper left portion of the screen and left click on **Return** as shown in Figure 70.

Figure 70

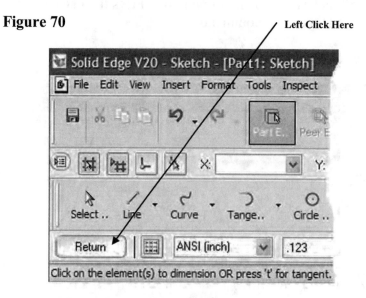

77. The screen should look similar to Figure 71.

Figure 71

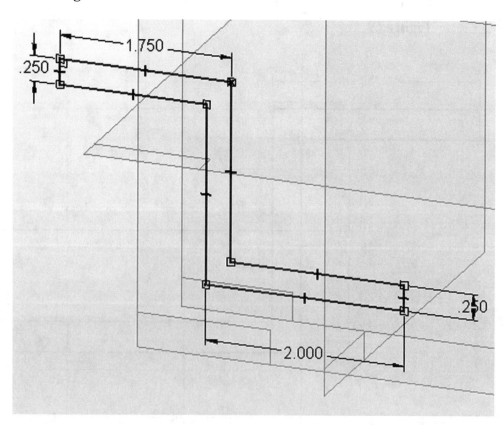

78. To utilize commands in the Model area, a sketch must be present and have no opens (non-connected lines). If there are any opens in the sketch an error message may appear.

79. Move the cursor to the middle left portion of the screen and left click on **Protrusion** as shown in Figure 72.

Figure 72

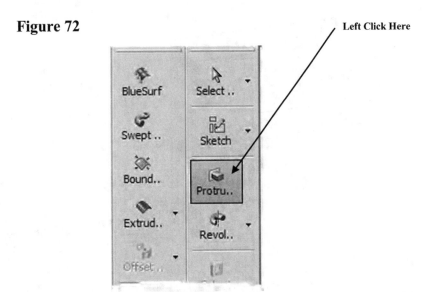

80. Move the cursor over any portion of the sketch causing the entire profile to turn red and left click as shown in Figure 73. The sketch will become yellow. After the sketch turns yellow, right click once.

Figure 73

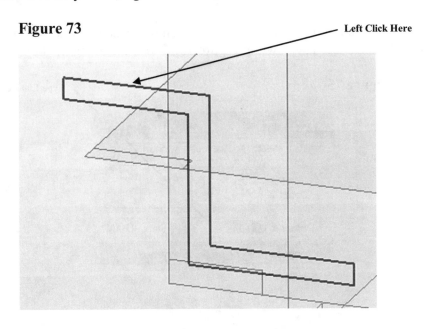

81. The sketch will become three dimensional as shown in Figure 74. The extruded surface will be attached to the cursor. Move the cursor forward and backward to confirm this.

Figure 74

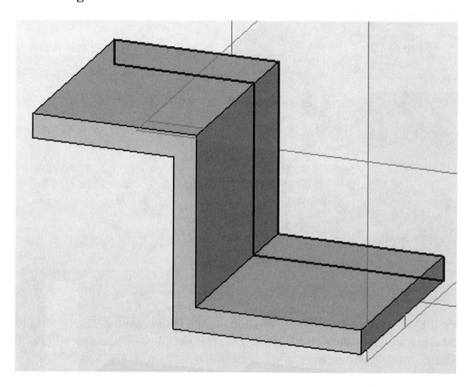

82. Move the cursor to the upper middle portion of the screen and type **1.00** next to the text "Distance" as shown in Figure 75. Press **Enter** on the keyboard.

Figure 75

Type 1.00 Here

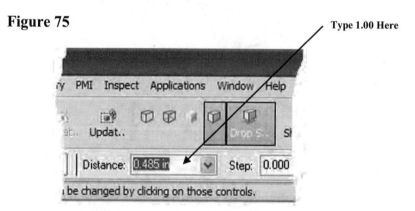

83. Left click anywhere around the sketch. Solid Edge will create a solid from the sketch as shown in Figure 76.

Figure 76

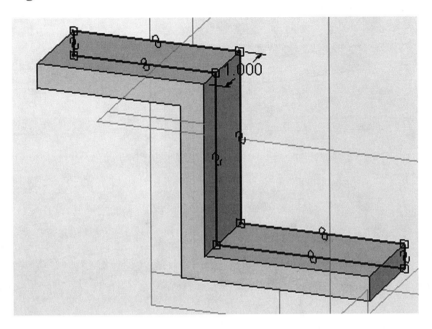

84. Move the cursor to the upper left portion of the screen and left click on **Finish** as shown in Figure 77.

Figure 77

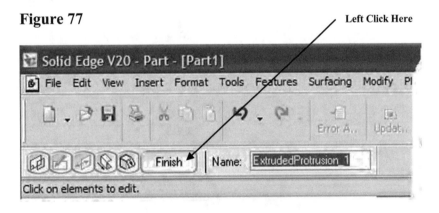

85.   The screen should look similar to Figure 78.

Figure 78

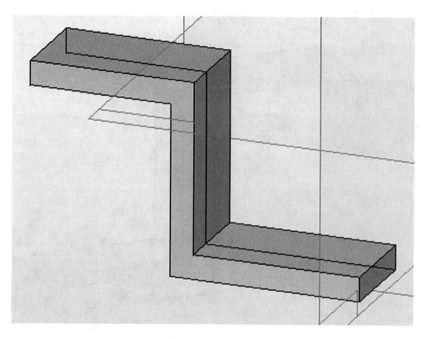

86.   Move the cursor to the middle left portion of the screen and left click on **Round** as shown in Figure 79.

Figure 79

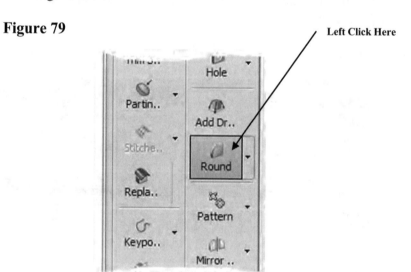

87. Move the cursor to the upper left portion of the screen and left click on the drop down arrow to the right of the text "Select". Left click on **Edge/Corner** as shown in Figure 80.

Figure 80

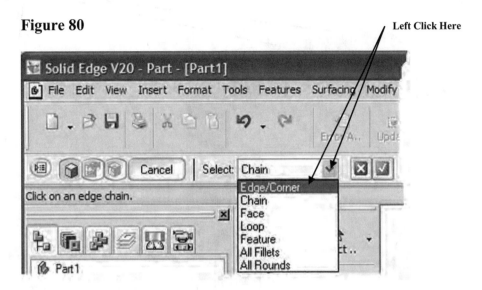

88. Move the cursor over the lower front edge causing it to turn red and left click once as shown in Figure 81. The edge will turn yellow.

Figure 81

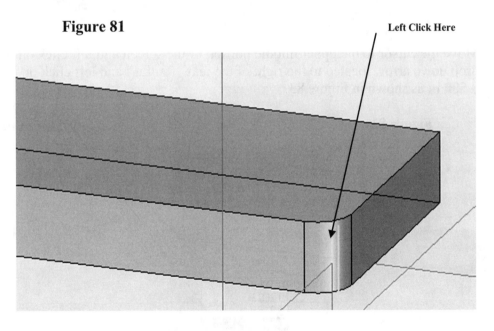

89. Left click on the remaining edges as shown in Figure 82.

Figure 82

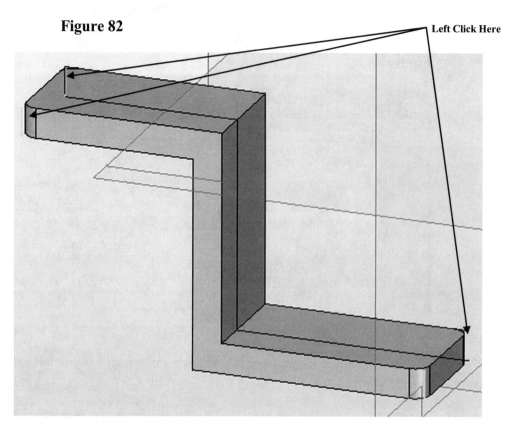

90. Move the cursor to the upper middle portion of the screen and left click on the drop down arrow located to the right of the text "Radius" and left click on **0.500 in** as shown in Figure 83.

Figure 83

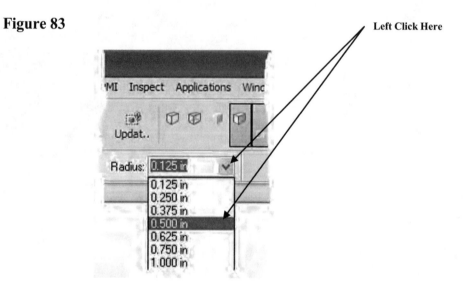

91. The screen should look similar to Figure 84.

Figure 84

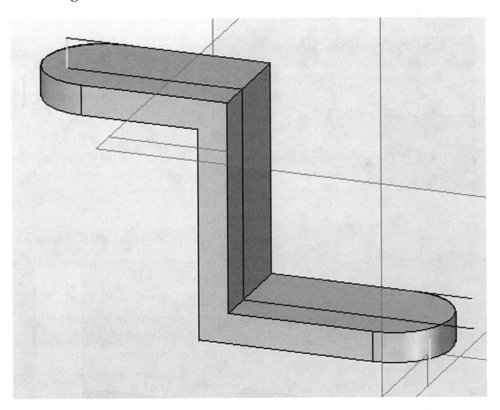

92. Move the cursor to the upper middle portion of the screen and left click on the green checkmark as shown in Figure 85.

Figure 85

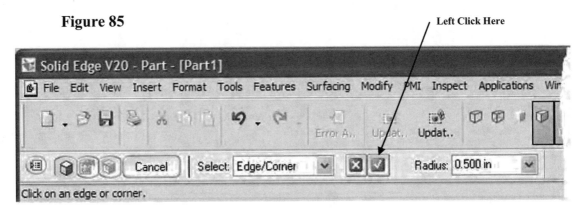

93. Move the cursor to the upper left portion of the screen and left click on **Preview** as shown in Figure 86.

Figure 86

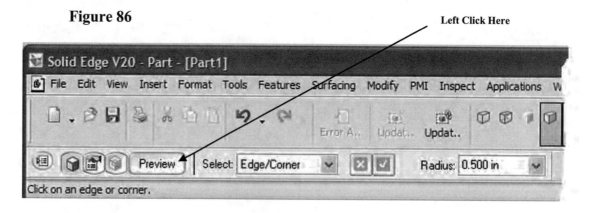

94. Move the cursor to the upper left portion of the screen and left click on **Finish** as shown in Figure 87.

Figure 87

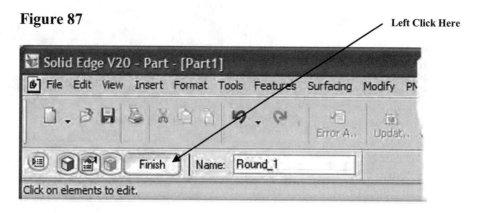

95. The screen should look similar to Figure 88.

Figure 88

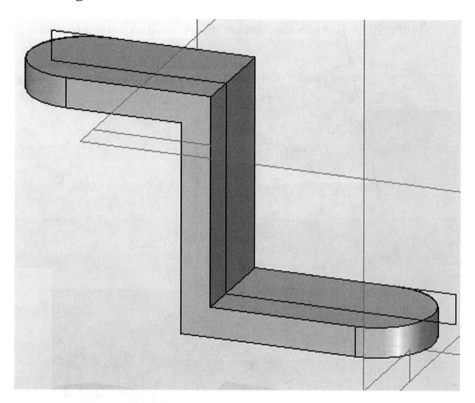

96. Move the cursor to the upper middle portion of the screen and left click on **Rotate** as shown in Figure 89.

Figure 89   Left Click Here

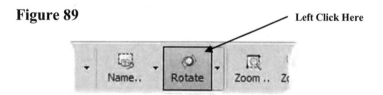

97. Left click on the left portion of the part (holding the left mouse button down). Drag the cursor to the right to view the upper right corner of the part as shown in Figure 90.

Figure 90   Left Click Here, Hold Left Mouse Button Down, and Drag to the Right

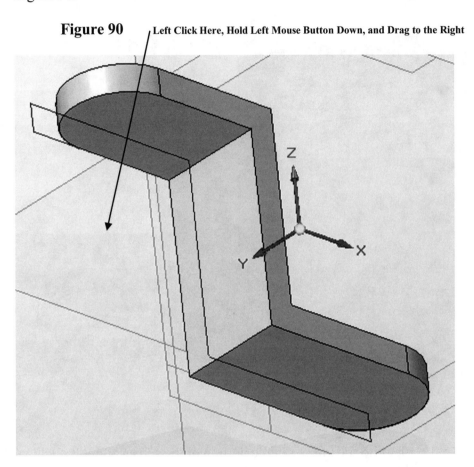

98. Move the cursor to the upper middle portion of the screen and left click on **View**. A drop down menu will appear. Left click on **Named Views** as shown in Figure 91.

Figure 91

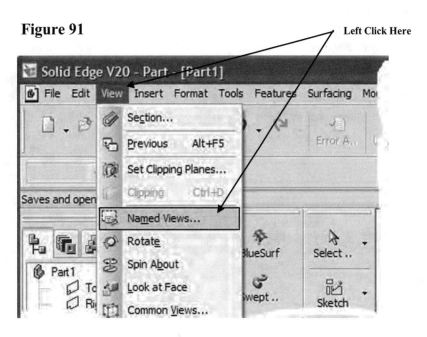

99. The Named Views dialog box will appear. Left click on **dimetric**. Left click on **Apply**. To edit the description of any view double click on the text. A cursor will appear allowing you to enter a user-defined description. Left click on **Close** as shown in Figure 92.

Figure 92

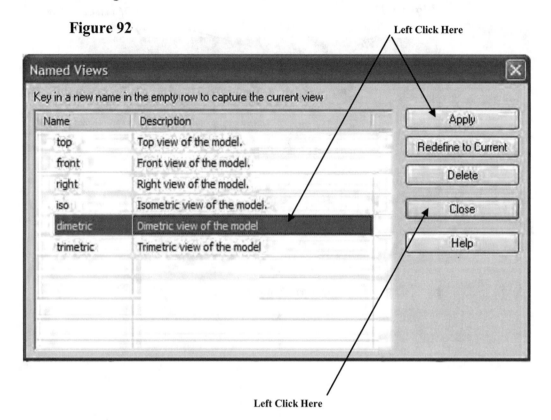

Left Click Here

Left Click Here

100. The screen should look similar to Figure 93.

Figure 93

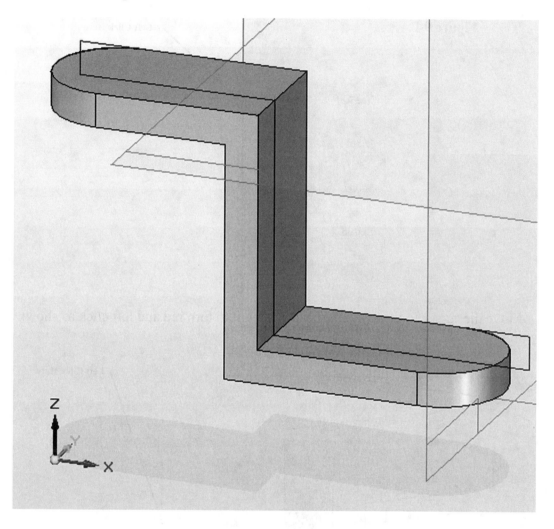

101. The next task will include cutting a hole in each of the ends. To do this, a sketch will need to be constructed on each surface. Move the cursor to the upper left portion of the screen and left click on **Sketch** as shown in Figure 94.

Figure 94

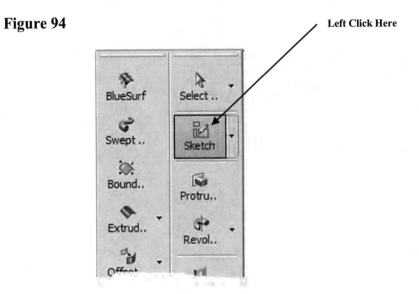

Left Click Here

102. Move the cursor to the lower surface causing it to turn red and left click as shown in Figure 95.

Figure 95

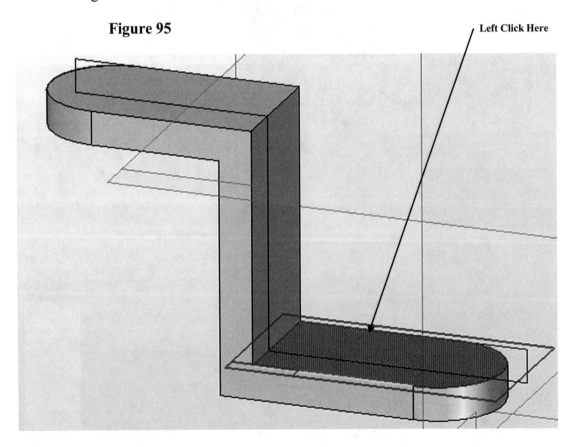

Left Click Here

103. Solid Edge will create a "sketch" on that particular surface. Solid Edge will also rotate the part around to provide a perpendicular view of the surface as shown in Figure 96. Notice the toolbar at the top of the screen has changed back to the sketch commands.

Figure 96

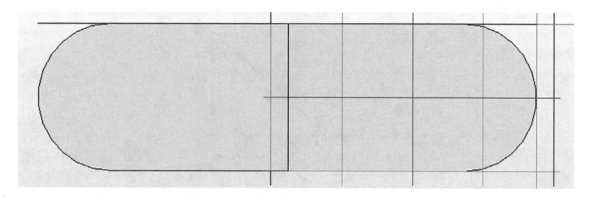

104. Move the cursor to the upper left portion of the screen and left click on **Circle by Center** as shown in Figure 97.

Figure 97

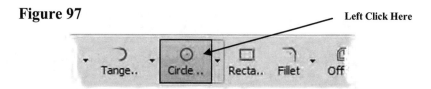

Left Click Here

105. Move the cursor to the edge of the part causing it to turn red. The center point of the circle will appear as shown in Figure 98.

Figure 98

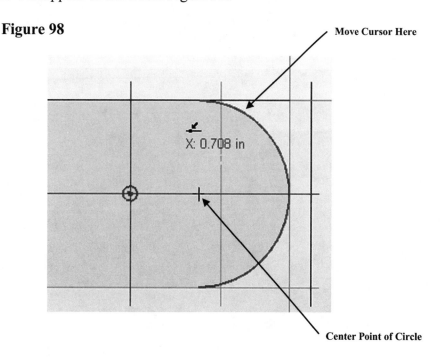

Move Cursor Here

Center Point of Circle

106. After the center marker appears, left click once. This will be the center of a circle which will later become a thru hole. Move the cursor out to the side to make the circle larger. Move the cursor out far enough to create a circle similar to the circle shown in Figure 99.

Figure 99

Left Click Here

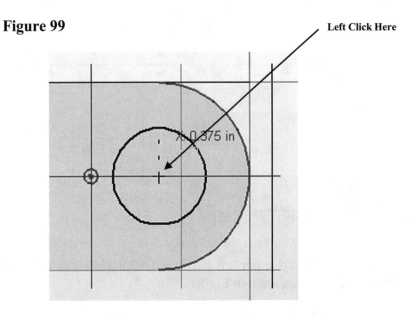

107. After the circle size looks similar to Figure 99, left click once.

108. Move the cursor to the upper middle portion of the screen and left click on **SmartDimension** as shown in Figure 100.

Figure 100

Left Click Here

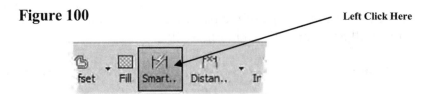

109. Move the cursor over the circle causing it to turn red and left click once as shown in Figure 101.

Figure 101

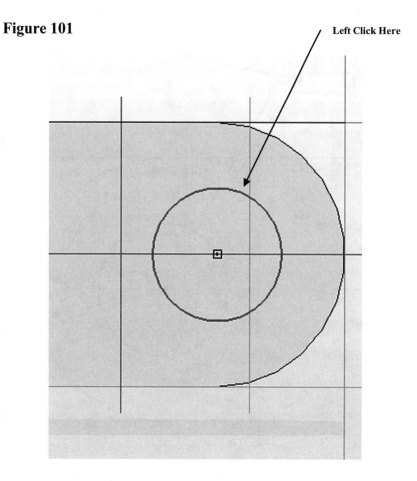

Left Click Here

110. The dimension box will appear and is attached to the cursor. Move the cursor to where the dimension will be placed and left click once as shown in Figure 102.

Figure 102

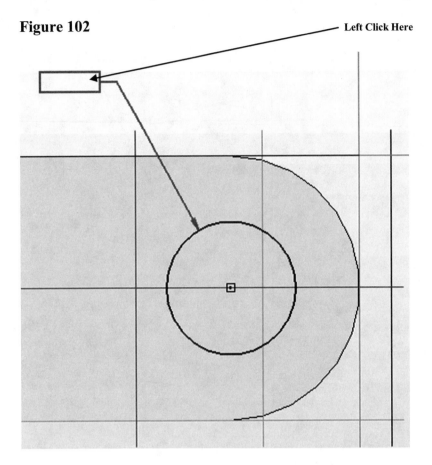

111. Move the cursor to the upper middle portion of the screen and type **.500** in the Dimension box as shown in Figure 103. Press **Enter** on the keyboard.

Figure 103

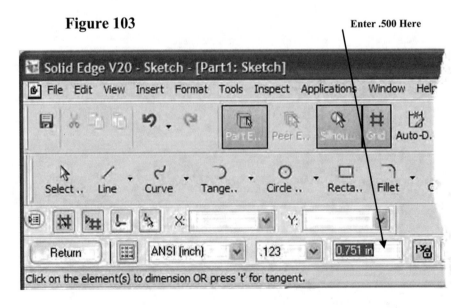

112. The diameter of the circle will become .500 inches as shown in Figure 104.

Figure 104

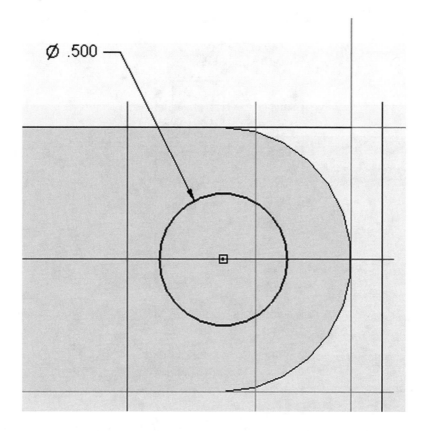

113. Move the cursor to the upper left portion of the screen and left click on **Return** as shown in Figure 105.

Figure 105

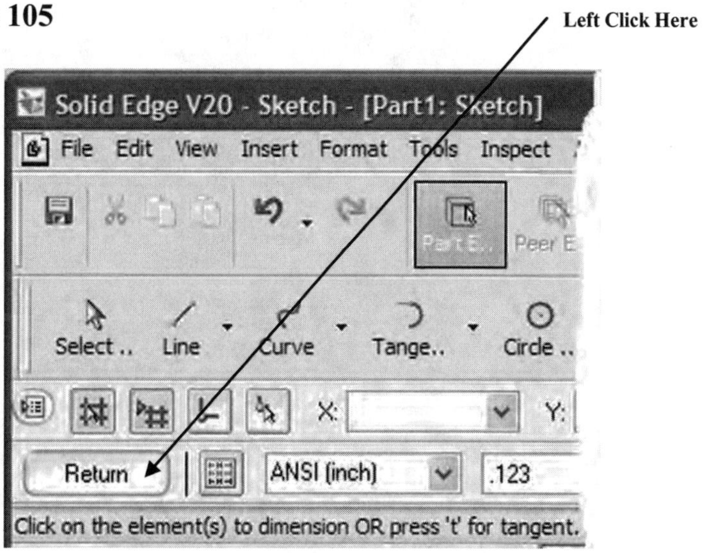

114. Solid Edge is now out of the Sketch command and into the Model command. Notice that the commands are now different as shown in Figure 106.

Figure 106

115. Move the cursor to the upper middle portion of the screen and left click on **Cutout** as shown in Figure 107.

Figure 107

Left Click Here

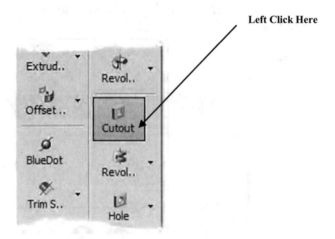

116. Move the cursor over the circle causing it to turn red and left click. Right click once causing the circle to turn yellow as shown in Figure 108.

Figure 108

Left Click/Right Click Here

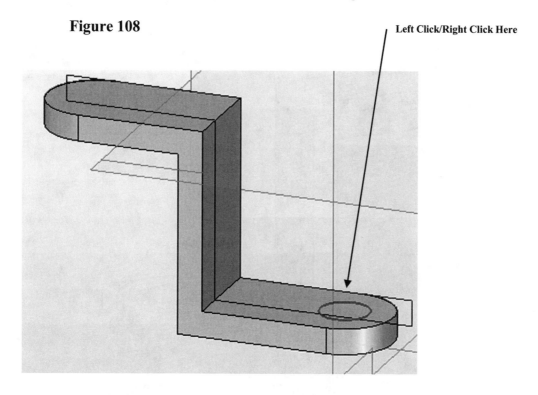

117. Move the cursor down below the bottom of the part. Type **.500** to the right of the text "Distance" as shown in Figure 109.

Figure 109

Type .500 Here

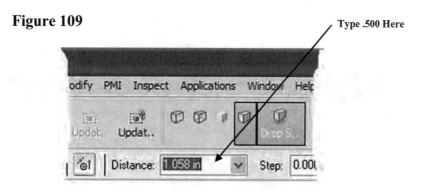

118. Left click anywhere around the sketch. A preview of the hole will be displayed as shown in Figure 110.

Figure 110

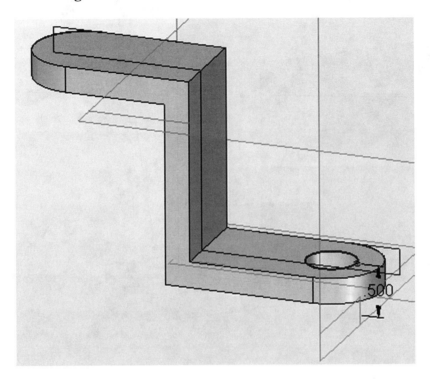

119. Move the cursor to the upper left portion of the screen and left click on **Finish** as shown in Figure 111.

Figure 111

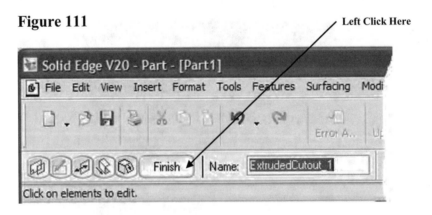

120. Use the Rotate command as detailed in instruction 96 to rotate the part upward. There should be a thru hole in the part similar to Figure 112.

Figure 112

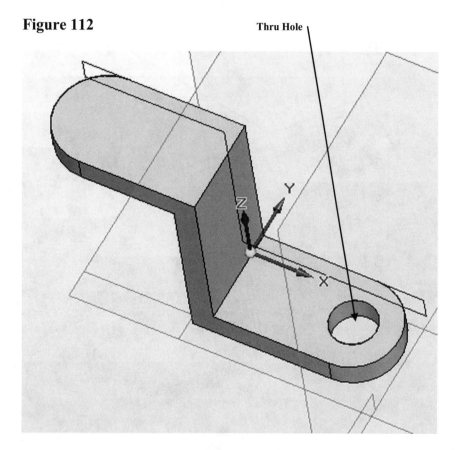

Thru Hole

121. Another method of creating a hole is to use the Hole command.

122. To use the Hole command a "Point" must be constructed in the Sketch area. Move the cursor to the upper middle portion of the screen and left click on **Sketch** as shown in Figure 113.

Figure 113

Left Click Here

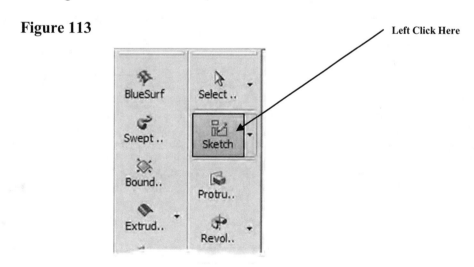

123. Move the cursor to the top portion of the part. The surface of the part will turn red. Left click as shown in Figure 114.

Figure 114

Left Click Here

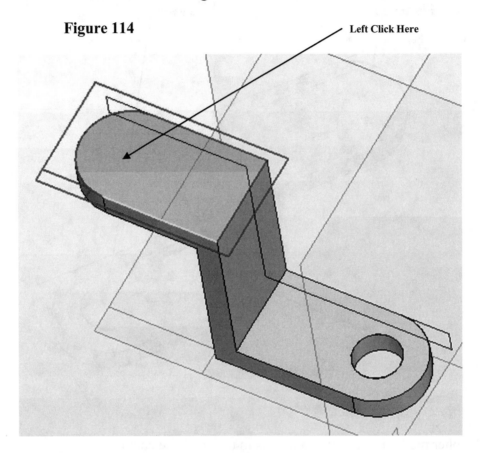

124. Solid Edge will return to the Sketch command as shown in Figure 115.

Figure 115

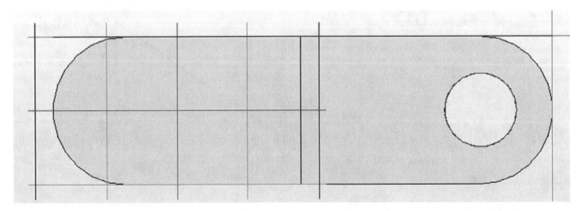

125. Move the cursor to the upper left portion of the screen and left click on the arrow next to the Line icon. A drop down menu will appear. Left click on **Point** as shown in Figure 116.

Figure 116

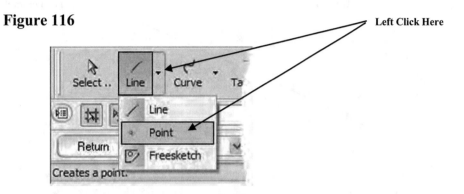

Left Click Here

126. Move the cursor to edge of the part. A portion of the edge will turn red. The center point of the edge radius will also appear. Left click once on the center point as shown in Figure 117.

Figure 117

Left Click Here

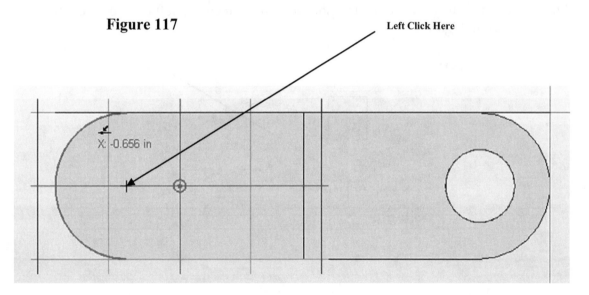

127. A small center marker will appear on the center of the edge radius as shown in Figure 118.

Figure 118

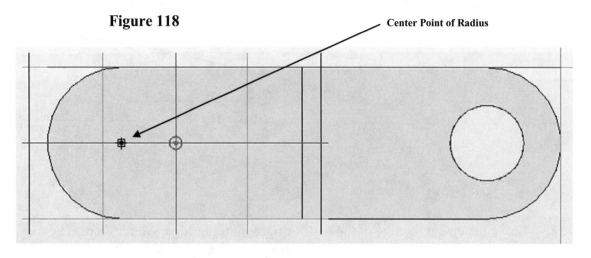

Center Point of Radius

128. Move the cursor the upper left portion of the screen and left click on **Return** as shown in Figure 119.

Figure 119

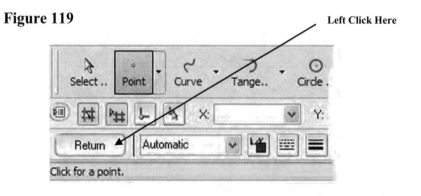

Left Click Here

129. Solid Edge is now out of the Sketch command and into the Model command. Notice that the commands at the top of the screen are now different. The screen should look similar to Figure 120.

Figure 120   Center Point of Edge Radius

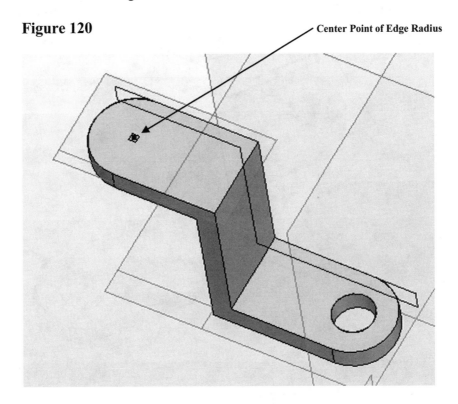

130. Move the cursor to the middle left portion of the screen and left click on **Hole** as shown in Figure 121.

Figure 121   Left Click Here

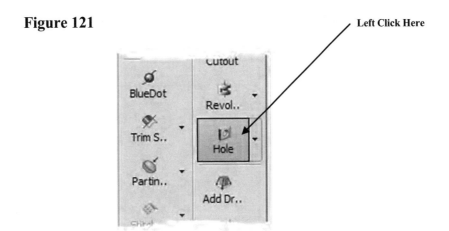

131. Left click on the Point that was just created in the Sketch command as shown in Figure 122.

Figure 122

Left Click Here

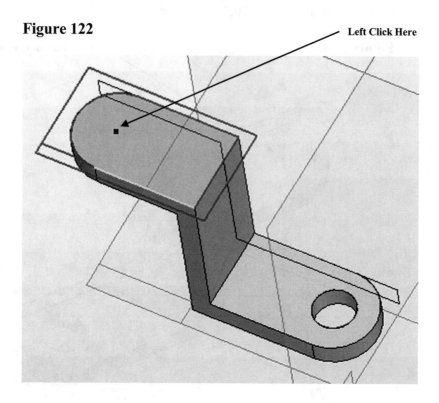

132. Solid Edge will return to the Sketch mode. A circle will be attached to the cursor. Move the circle over the Point and left click as shown in Figure 123.

Figure 123

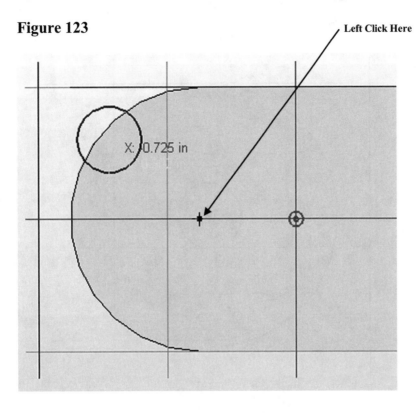

133.   After the center of the circle is located on the Point, press the **Esc** key on the keyboard. The screen should look similar to Figure 124.

Figure 124

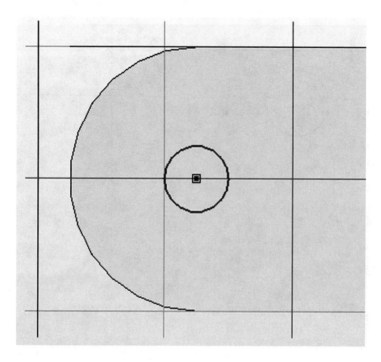

134.   Move the cursor to the upper left portion of the screen and left click on **Return** as shown in Figure 125.

Figure 125

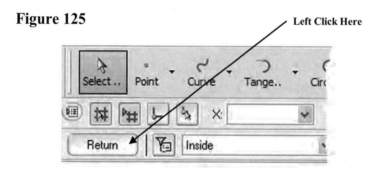

135. Solid Edge will return to the Model mode. A hole with an arrow will appear as shown in Figure 126.

Figure 126   Hole with Red Arrow

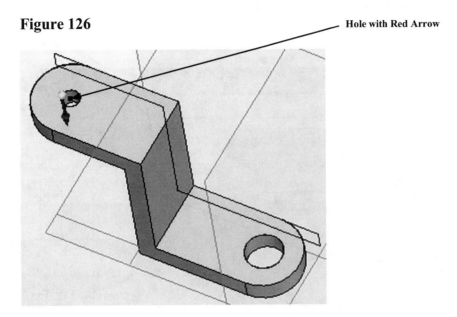

136. Move the cursor to the upper left portion of the screen and left click on the **Hole Options** icon as shown in Figure 127.

Figure 127   Left Click Here

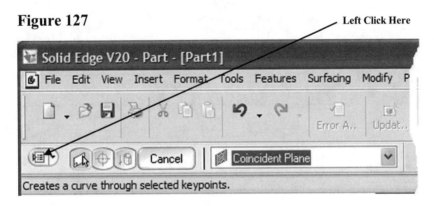

137. The **Hole Options** dialog box will appear. Left click on the drop down arrow next to the text "Type". A drop down menu will appear. Left click on **Counterbore** as shown in Figure 128.

Figure 128

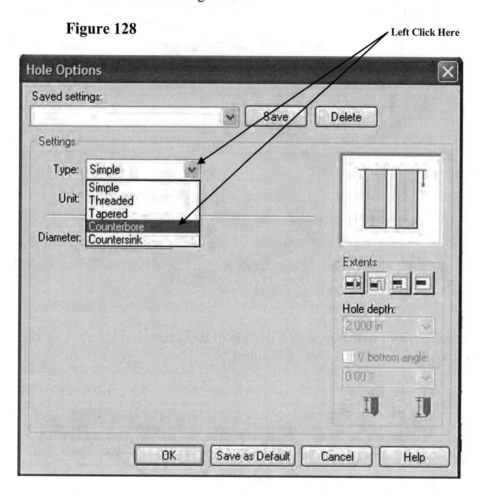

138.   More options will appear in the Hole Options dialog box. Enter **.25** for Diameter, **.500** for Counterbore diameter and **.125** for the Counterbore depth as shown in Figure 129.

Figure 129

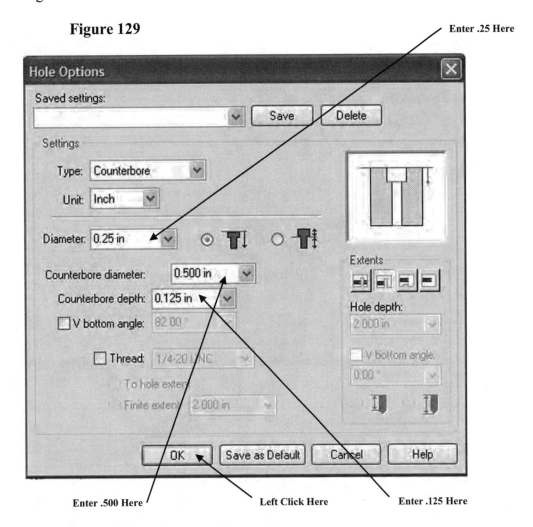

139.   Left click on **OK** as shown in Figure 129.

140.   Move the cursor to the upper left portion of the screen and left click on **Finish** as shown in Figure 130.

Figure 130

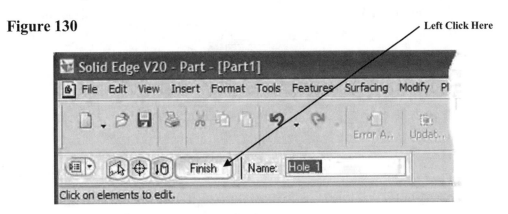

141. The screen should look similar to Figure 131.

Figure 131

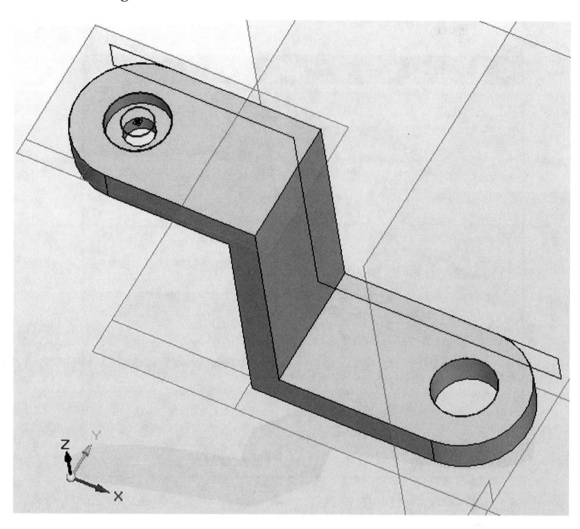

142. To ensure that the hole is correct move the cursor to the upper right portion of the screen and left click on the **Rotate** icon as shown in Figure 132.

Figure 132

Left Click Here

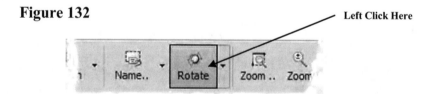

143. Left click anywhere around the part and hold the left mouse button down. Drag the cursor upward. The part will rotate upward as shown in Figure 133.

Figure 133

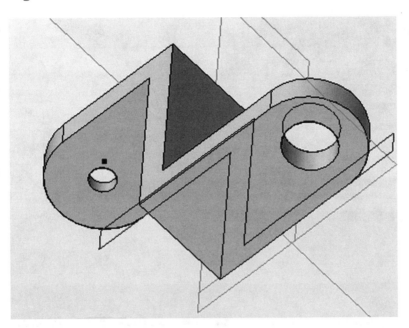

144. Holding the left mouse button down keeps the part attached to the cursor. To view the part in a standard view, left click on the **Name** command as shown in Figure 134.

Figure 134   Left Click Here

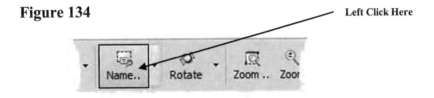

145. The Named Views dialog box will appear. Left click on **dimetric**. Left click on **Apply** as shown in Figure 135.

Figure 135

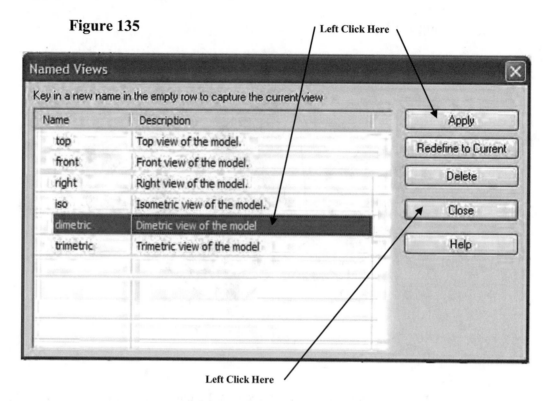

Left Click Here

Left Click Here

146. Left click on **Close**.

147. The screen should look similar to Figure 136.

Figure 136

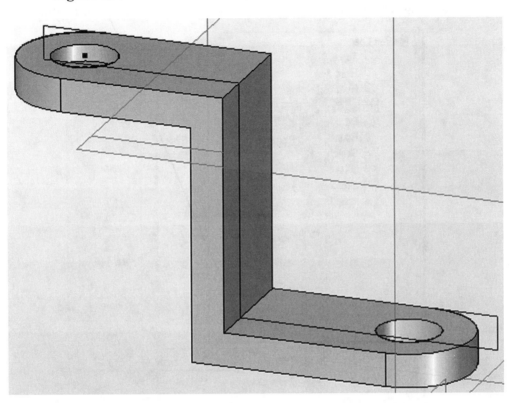

148. Move the cursor over the text "Sketch 1" in the path window causing a red box to appear as shown in Figure 137.

Figure 137   Red Box

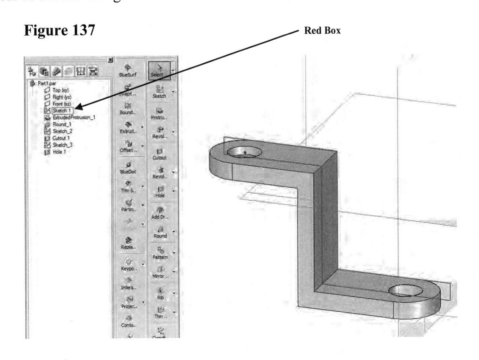

149. After the red box appears right click once. A pop up menu will appear. Left click on **Hide** as shown in Figure 138.

Figure 138

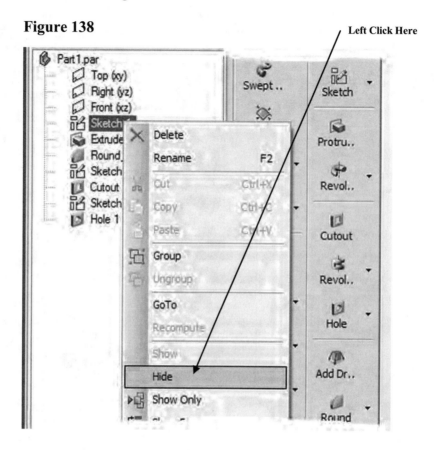

150. The original sketch used to create the part will now be hidden from view. The screen should look similar to Figure 139.

Figure 139

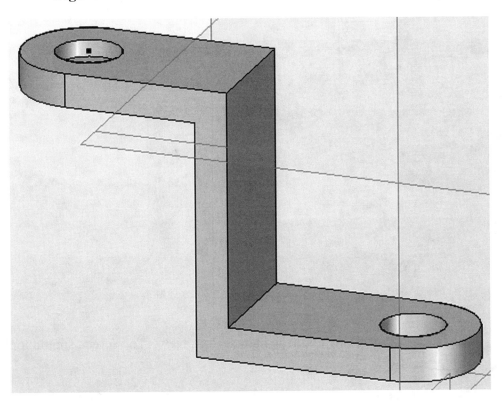

151. Move the cursor over the text "Front (xz)" in the path window as shown in Figure 140. A red box will appear around the text.

Figure 140

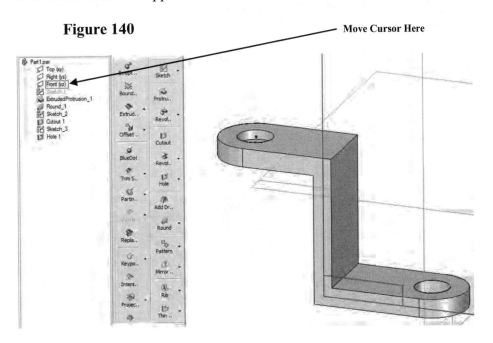

152. After the red box appears around the text, right click once. A pop up menu will appear. Left click on **Hide** as shown in Figure 141.

Figure 141

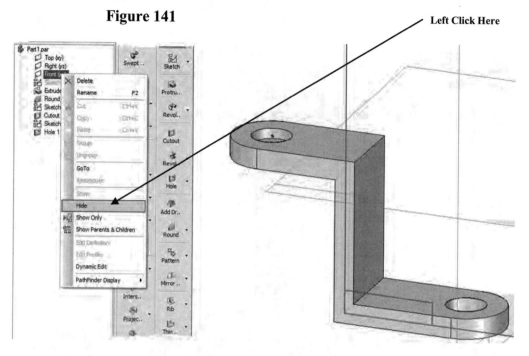

153. The Front (xz) plane will now be hidden from view. The screen should look similar to Figure 142.

Figure 142

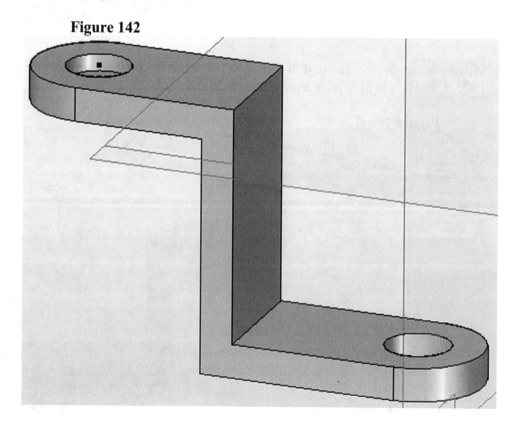

154. Repeat the previous steps to hide the Right (yz) and Top (xy) planes. After this is complete, the screen should look similar to Figure 143.

Figure 143

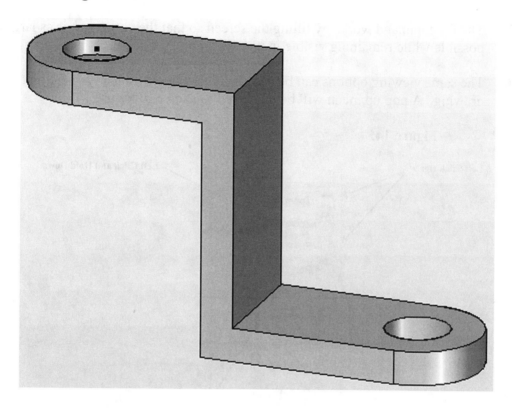

155. Other commands for viewing are located at the top of the screen as shown in Figure 144.

Figure 144

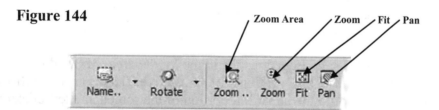

156. The Zoom Area command works by using the cursor to draw a window around the area you want to zoom in on. After selecting the "Zoom Area" icon, hold the left mouse button down and drag a diagonal box around the desired area. Release the left mouse button when the proper amount of zoom is achieved.

157. The Zoom command works similar to the Zoom Area command. Start by selecting the "Zoom" icon. Left click on the drawing and hold the left mouse button down while dragging the cursor up and down until the proper amount of zoom is achieved.

158. The Pan command works similar to the Zoom command. Start by selecting the "Pan" icon. Left click on the drawing and hold the left mouse button down while moving the cursor up and down or side to side. Release the mouse button after the desired view is achieved.

159. The Fit command works by filling the screen so that the entire part is as large as possible while remaining visible in its entirety.

160. The same viewing options can be accessed by right clicking anywhere on the drawing. A pop up menu will be displayed as shown in Figure 145.

Figure 145

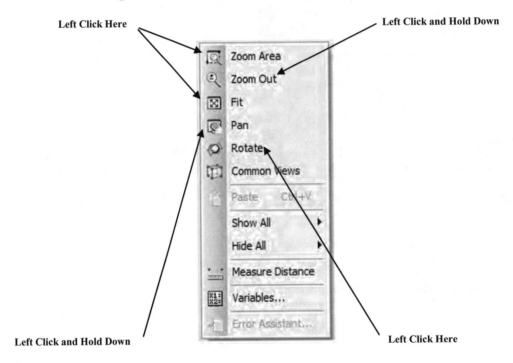

Left Click Here

Left Click and Hold Down

Left Click and Hold Down

Left Click Here

161. To save a file, Solid Edge must be in the Model area. If Solid Edge is in the Sketch area, the "Close" option will be inactive. To move to the Model area, move the cursor to the upper left portion of the screen and left click on **Return** as described in instruction number 76.

Drawing Activities

Problem 1

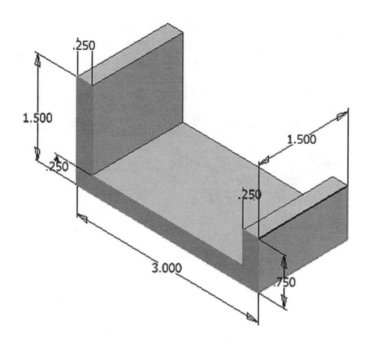

Problem 2

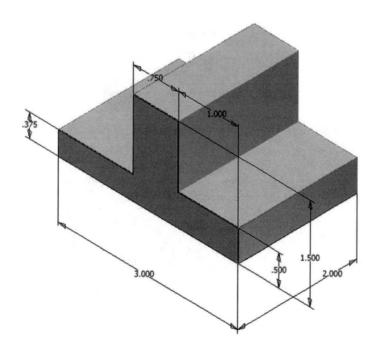

Problem 3

Extrude Center Section .25 Deep

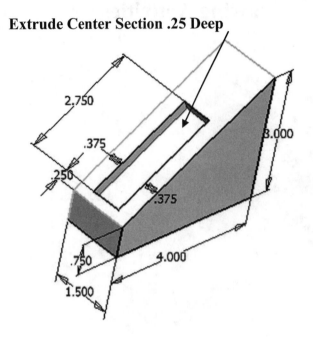

Problem 4

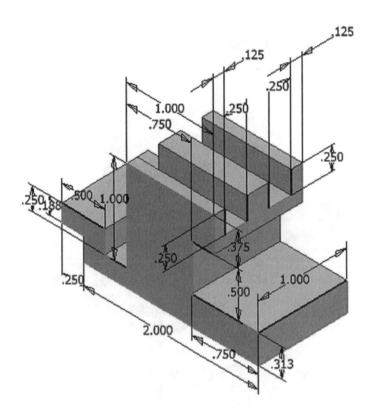

Problem 5

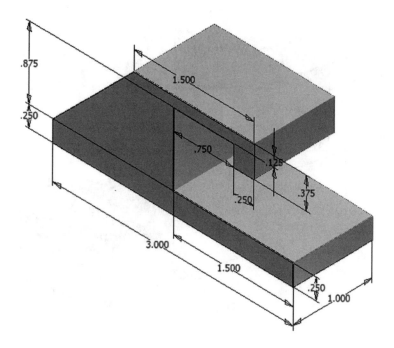

Problem 6

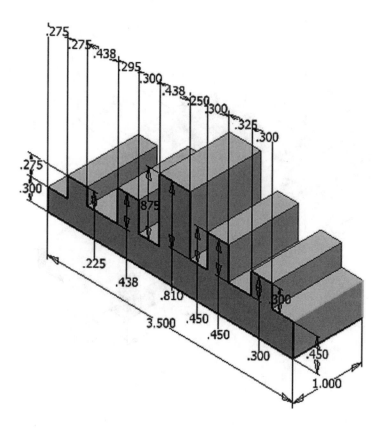

Problem 7

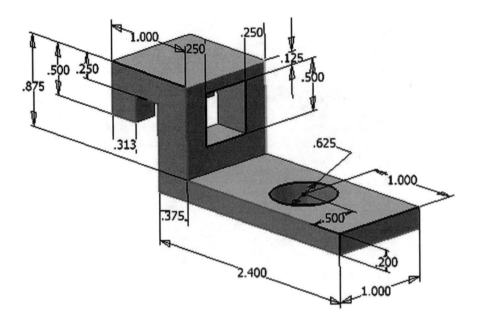

Problem 8

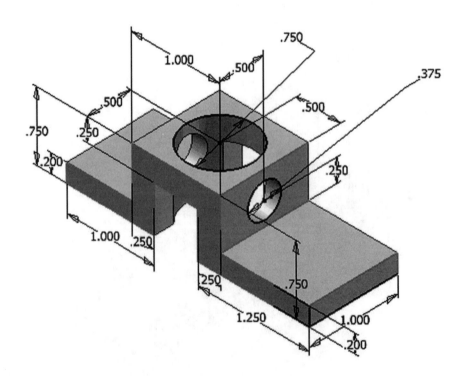

Chapter 2 Learning More Basics

Objectives:

- Create a simple sketch using the Sketch command
- Dimension a sketch using the SmartDimension command
- Revolve a sketch using the Revolve command
- Create a hole using the Cutout command
- Create a series of holes using the Circular Pattern command

Chapter 2 includes instruction on how to design the part shown below.

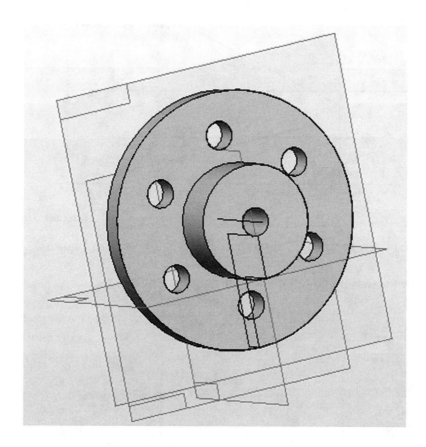

1. Start Solid Edge by referring to "Chapter 1 Getting Started".

2. After Solid Edge is running, begin a new sketch.

3. Move the cursor to the upper left corner of the screen and left click on **Line** as shown in Figure 1.

Figure 1

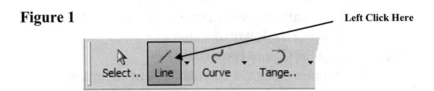

Left Click Here

4. Move the cursor to the lower left portion of the screen and left click once. This will be the beginning end point of a line as shown in Figure 2.

Figure 2

Beginning Endpoint of Line

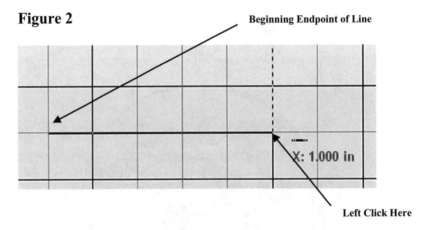

X: 1.000 in

Left Click Here

5. Move the cursor to the right and left click once as shown in Figure 2.

6. Move the cursor up and left click once as shown in Figure 3.

Figure 3

Left Click Here

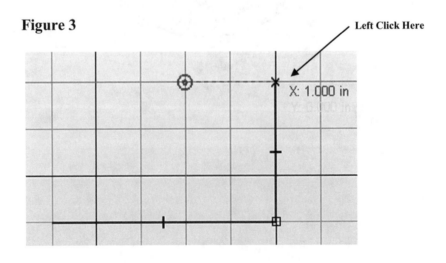

X: 1.000 in

7.   Move the cursor to the right and left click once as shown in Figure 4.

Figure 4

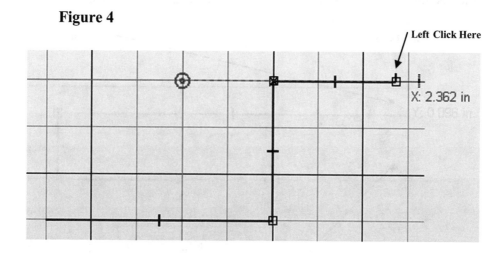

8.   Move the cursor up and left click once as shown in Figure 5.

Figure 5

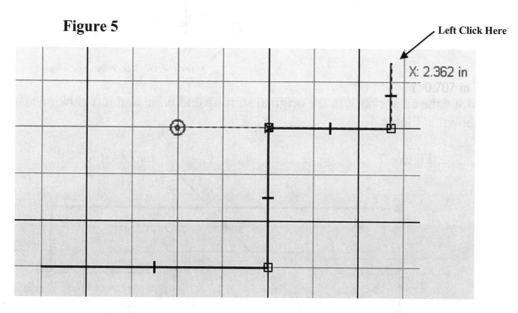

9.   Move the cursor to the left.   Ensure that the dots between the first end point and the last end point appear as shown in Figure 6. Left click once.

Figure 6

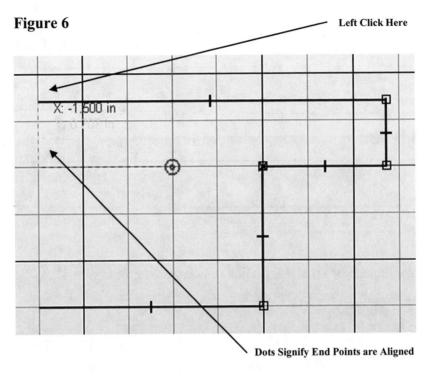

10.   Move the cursor back to the original starting end point and left click once as shown in Figure 7.

Figure 7

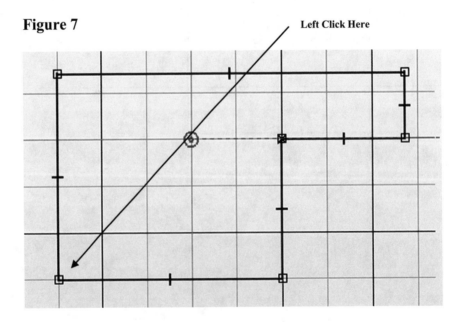

11.   Move the cursor to the middle left portion of the screen and left click on
       SmartDimension as shown in Figure 8.

Figure 8   Left Click Here

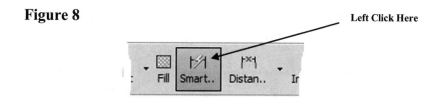

12.   After selecting **SmartDimension** move the cursor over the bottom horizontal line
       until it turns red as shown in Figure 9. Select the line by left clicking anywhere
       on the line **or** on each of the end points. To use the end points of the line, move
       the cursor over one of the end points. Left click once and move the cursor to the
       other end point and left click again. The dimension box will be attached to the
       cursor.

Figure 9   Turned Red

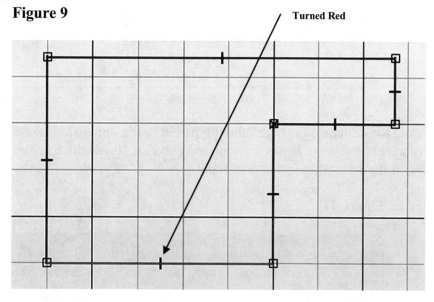

13. Move the cursor to where the dimension will be placed and left click once as shown in Figure 10.

Figure 10

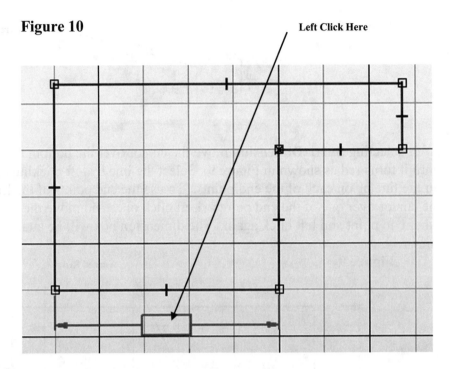

Left Click Here

14. While the dimension is highlighted type **.50** in the dimension box as shown in Figure 11. Press the **Enter** key on the keyboard. Press the **Esc** key once or twice to exit the SmartDimension command.

Figure 11

Enter .50 Here

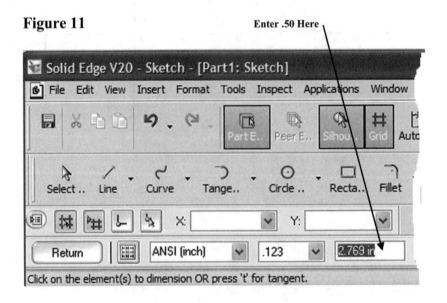

15. The dimension of the line will become .50 inches as shown in Figure 12.

Figure 12

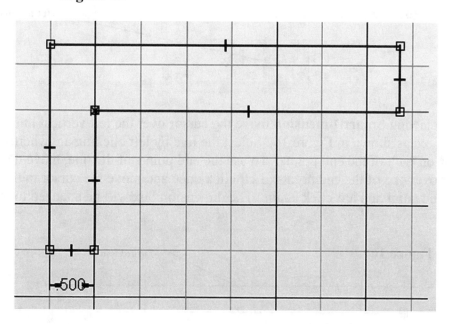

16. To view the entire drawing move the cursor to the upper middle portion of the screen and left click once on the "Fit" icon as shown in Figure 13.

Figure 13   Left Click Here

17. The drawing will "fill up" the entire screen. If the drawing is still too large, left click on the "Zoom" icon as shown in Figure 14. After selecting the icon, hold the left mouse button down and move the cursor up and down to achieve the desired view of the sketch.

Figure 14   Left Click Here

95

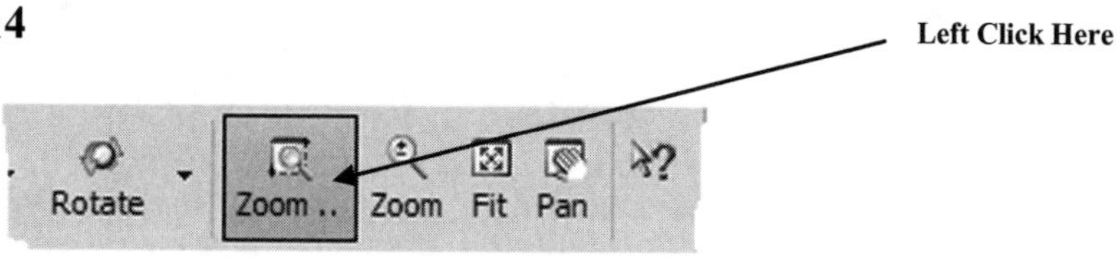

18.   Move the cursor to the middle left portion of the screen and left click on **SmartDimension** as shown in Figure 15.

Figure 15   Left Click Here

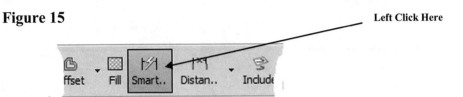

19.   After selecting **SmartDimension** move the cursor over the left vertical line until it turns red as shown in Figure 16. Select the line by left clicking anywhere on the line **or** on each of the end points. To use the end points of the line, move the cursor over one of the end points. Left click once and move the cursor to the other end point and left click again. The dimension box will be attached to the cursor.

Figure 16   Turned Red

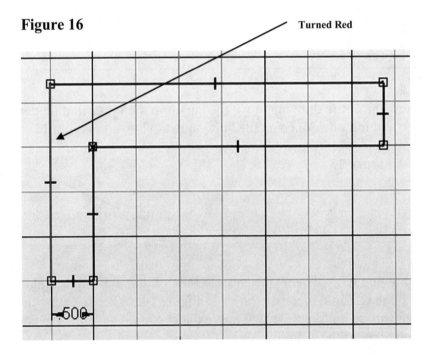

20. Move the cursor to where the dimension will be placed and left click once as shown in Figure 17.

Figure 17

Left Click Here

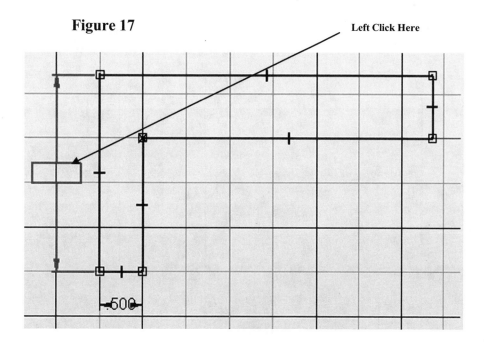

21. While the dimension is highlighted type **2.0** in the dimension box as shown in Figure 18. Press the **Enter** key on the keyboard.

Figure 18   Enter 2.0 Here

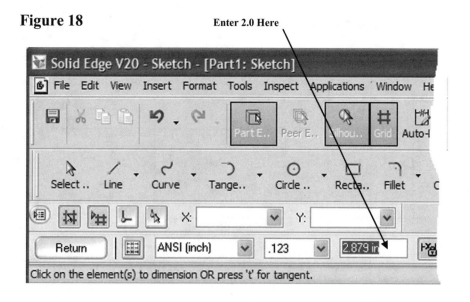

22. The dimension of the line will become 2.0 inches as shown in Figure 19. Press the **Esc** key once or twice to exit the SmartDimension command. Use the Zoom icons to zoom out if necessary.

Figure 19

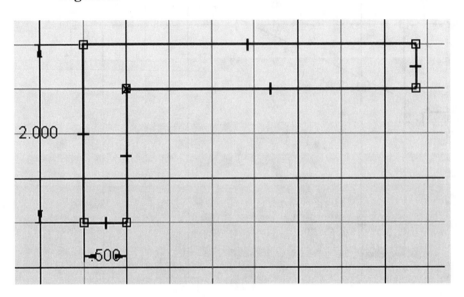

23. Select the next line by left clicking anywhere on the line **or** on each of the end points. The dimension box will be attached to the cursor.

Figure 20

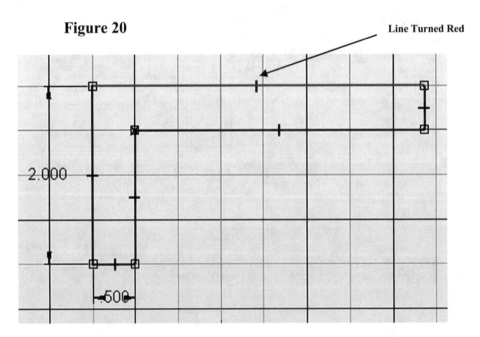

24. Move the cursor to where the dimension will be placed and left click once as shown in Figure 21.

Figure 21

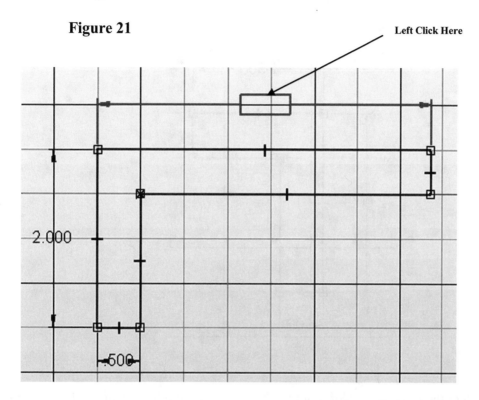

25. While the dimension is highlighted type **1.5** in the dimension box as shown in Figure 22. Press the **Enter** key on the keyboard. Press the **Esc** key once or twice to exit the SmartDimension command.

Figure 22

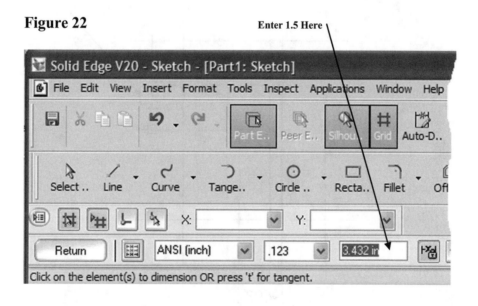

26.   The dimension of the line will become **1.5** inches as shown in Figure 23. Use the Zoom icons to zoom out if necessary.

Figure 23

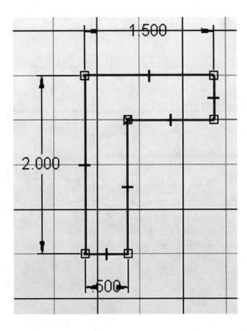

27.   Select the next line by left clicking anywhere on the line **or** on each of the end points. The line will turn red as shown in Figure 24. The dimension box will be attached to the cursor.

Figure 24

Turned Red

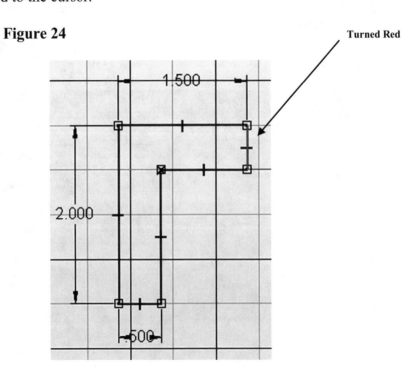

28. Move the cursor to where the dimension will be placed and left click once as shown in Figure 25.

Figure 25

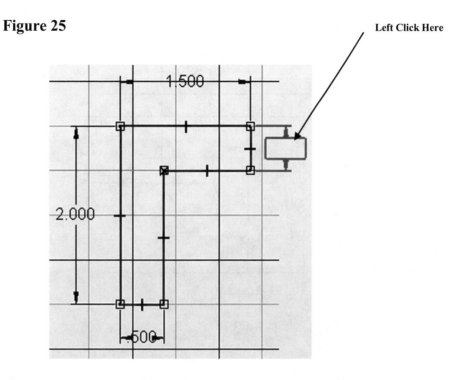

Left Click Here

29. While the dimension is highlighted type **.75** in the dimension box as shown in Figure 26. Press the **Enter** key on the keyboard. Press the **Esc** key once or twice to exit the SmartDimension command.

Figure 26

Enter .75 Here

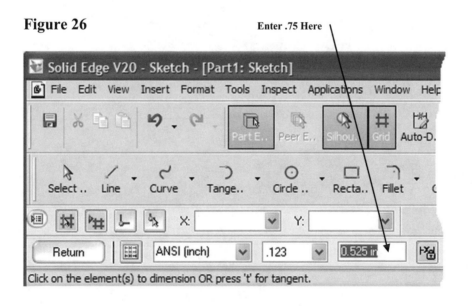

30.   The dimension of the line will become **.75** inches as shown in Figure 27. Use the Zoom icons to zoom out if necessary.

Figure 27

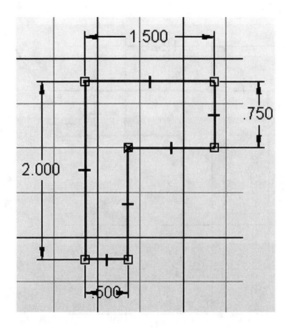

31.   Move the cursor to the upper left corner of the screen and left click on **Line** as shown in Figure 28.

Figure 28

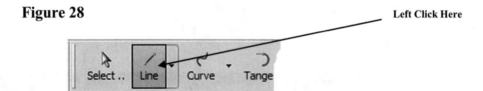

32.   Draw a line parallel to the top horizontal line as shown in Figure 29.

Figure 29

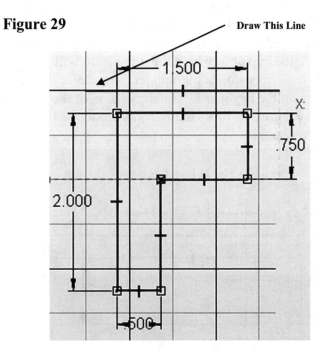

33. Move the cursor to the upper middle portion of the screen and left click on **SmartDimension.** Move the cursor over the parallel line that was just drawn causing it to turn red. Left click once. Move the cursor over the top line of the part causing it to turn red and left click once. The dimension box will be attached to the cursor. Move the cursor to where the dimension will be placed and left click once. While the dimension is highlighted type **.25** in the dimension box and press the **Enter** key on the keyboard. The dimension of the line will become .25 inches as shown in Figure 30. Press the **Esc** key a once or twice to exit the SmartDimension command.

Figure 30   Dimension Line .25 From Line Below

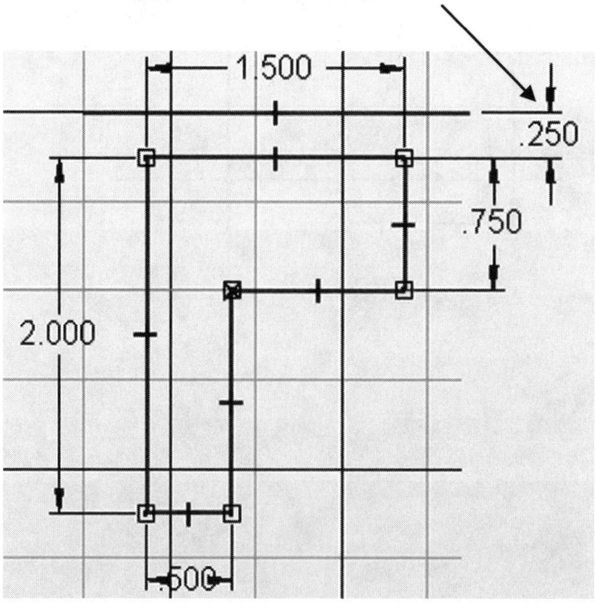

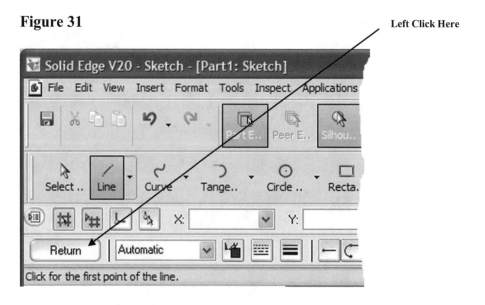

34. Move the cursor to the upper left portion of the screen and left click on **Return** as shown in Figure 31.

Figure 31   Left Click Here

35. After the sketch is complete it is time to revolve the sketch into a solid.

36. Solid Edge is now out of the Sketch area and into the Model area. Notice that the commands at the top of the screen are now different. To work in the Model area a sketch must be present and have no opens (non-connected lines). If there are any opens in the sketch an error message may appear. Your screen should look similar to Figure 32.

Figure 32

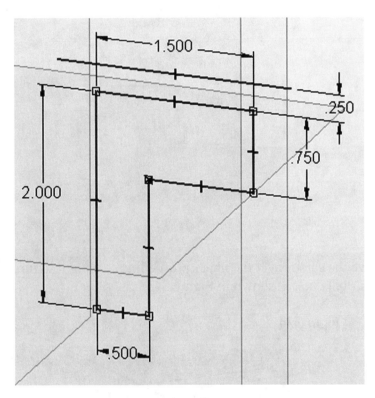

37.   Move the cursor to the upper left portion of the screen and left click on **Revolved Protrusion** as shown in Figure 33.

Figure 33

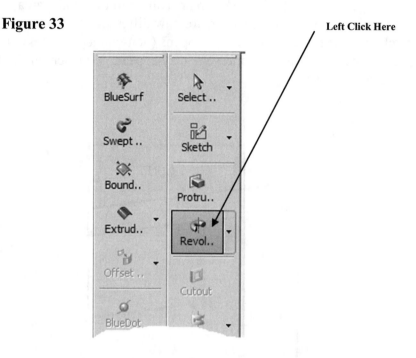

Left Click Here

38.   Move the cursor over the upper profile line causing it to turn red. Left click once then right click as shown in Figure 34.

Figure 34

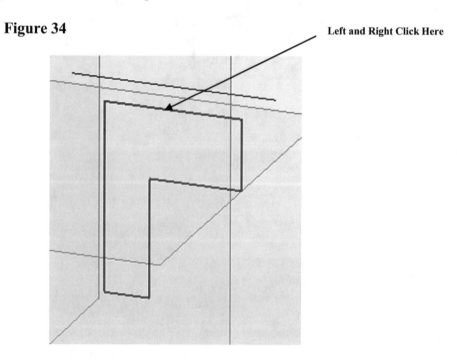

Left and Right Click Here

39. Move the cursor over the "axis" line that was drawn above the part as shown in Figure 35. The line will turn red.

Figure 35

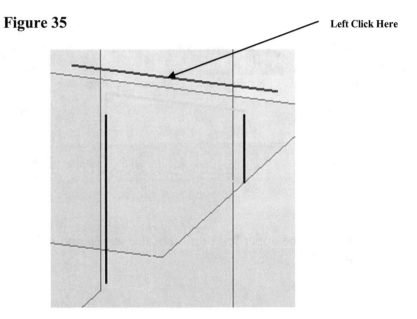

Left Click Here

40. Left click anywhere around the sketch. A preview of the revolved protrusion will be attached to the cursor and appear as shown in Figure 36.

Figure 36

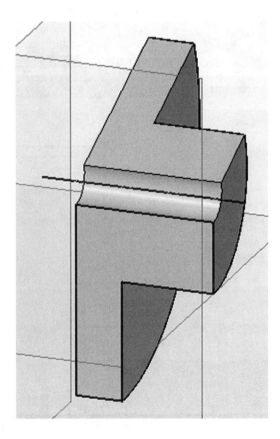

41.   Move the cursor to the upper middle portion of the screen and type **360** in the box to the right of the text "Angle" as shown in Figure 37. Press the **Enter** key on the keyboard.

Figure 37

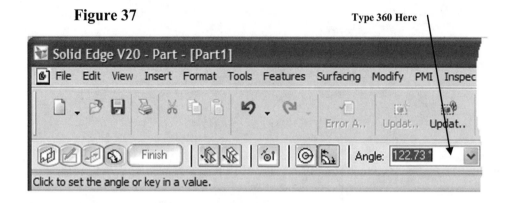

42.   Move the cursor to the upper left portion of the screen and left click on **Finish** as shown in Figure 38.

Figure 38

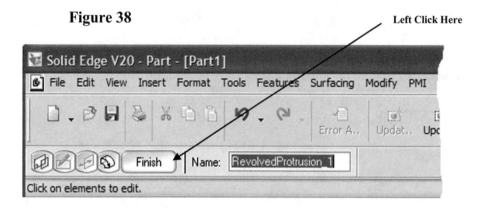

43. Your screen should look similar to Figure 39. You may have to zoom out to view the entire part.

Figure 39

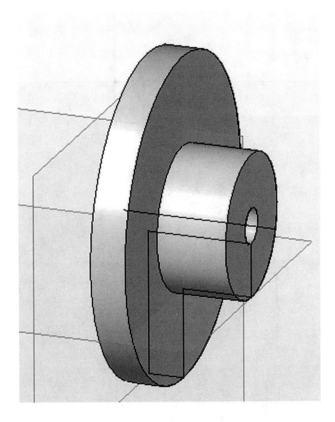

44. Move the cursor to the upper left portion of the screen and left click on the **Sketch** command as shown in Figure 40.

Figure 40

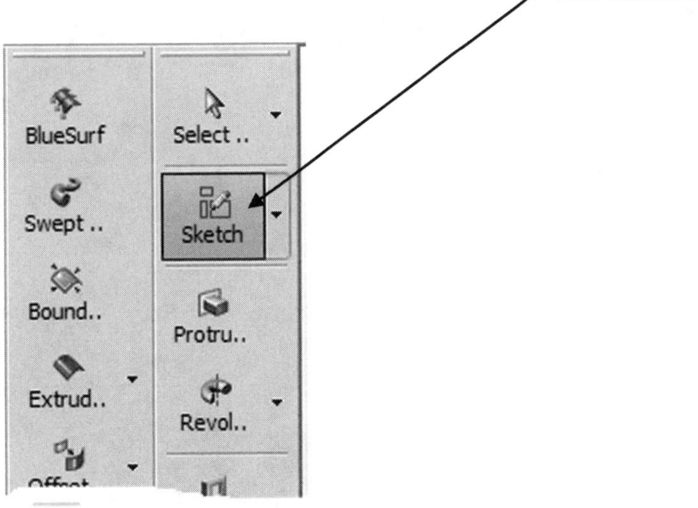

Left Click Here

109

45. Move the cursor to the edge of the part causing the edges to turn red. After the edges become red, left click on the surface as shown in Figure 41.

Figure 41

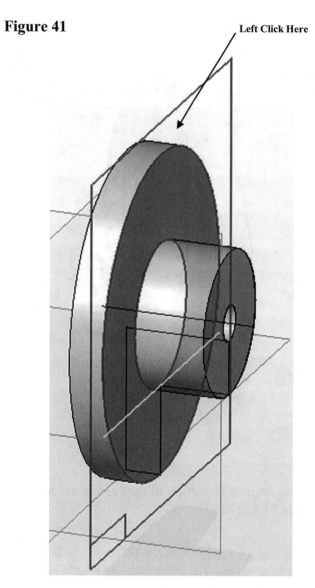

Left Click Here

46.   Solid Edge will begin a new sketch on the selected surface. Your screen should look similar to Figure 42.

Figure 42

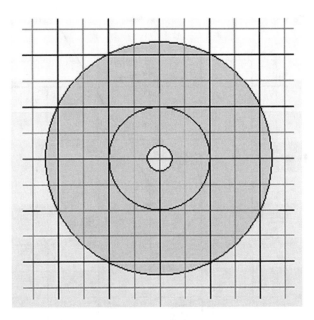

47.   Move the cursor to the upper left corner of the screen and left click on **Line** as shown in Figure 43.

Figure 43   Left Click Here

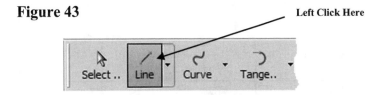

48.   Move the cursor over the small circle and wait a few seconds. A pair of concentric circles will appear in the center of the part. Left click in the center of the circles as shown in Figure 44.

Figure 44

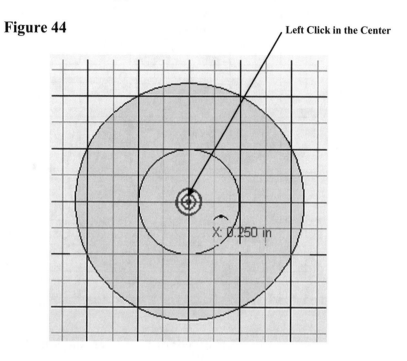

49.   Move the cursor straight up and left click as shown in Figure 45.

Figure 45

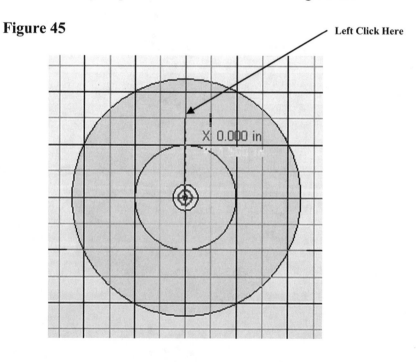

50. Right click as shown in Figure 46.

Figure 46

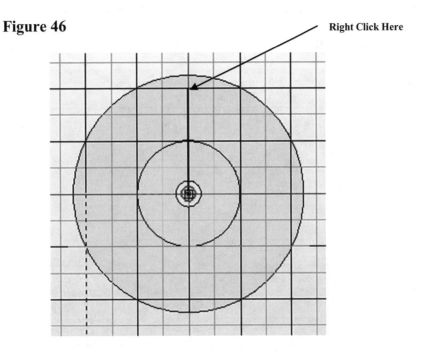

Right Click Here

51. Move the cursor to the upper left portion of the screen and left click on **SmartDimension** as shown in Figure 47.

Figure 47

Left Click Here

52. Move the cursor over the line that was just drawn. The line will turn red as shown in Figure 48. Left click once. The dimension box will be attached to the cursor.

Figure 48

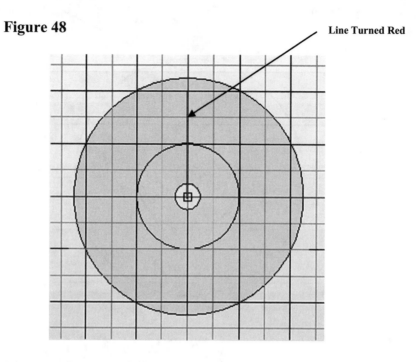

Line Turned Red

53. Move the cursor to where the dimension will be placed and left click once as shown in Figure 49.

Figure 49

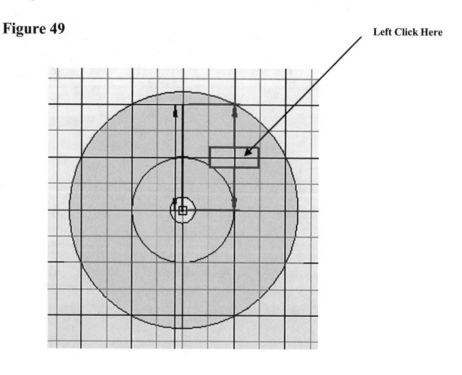

Left Click Here

54. To edit the dimension, type **1.5** in the Dimension box (while the current dimension is highlighted) as shown in Figure 50. Press **Enter** on the keyboard.

Figure 50   Type 1.5 Here

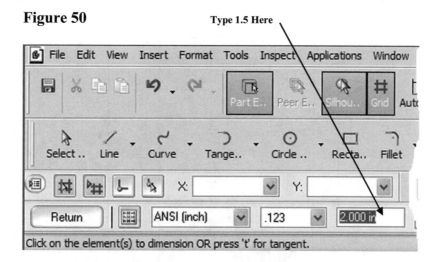

55. Right click anywhere on the screen. The dimension of the line will become 1.5 inches as shown in Figure 51. Use the Zoom icons to zoom out if necessary.

Figure 51

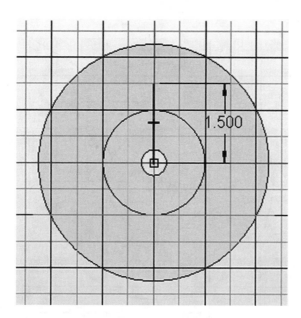

56. Move the cursor to the upper left portion of the screen and left click on **Circle by Center** as shown in Figure 52.

Figure 52   Left Click Here

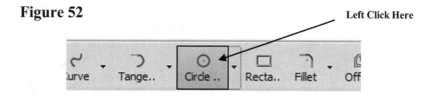

57. Left click on the endpoint of the line as shown in Figure 53.

Figure 53

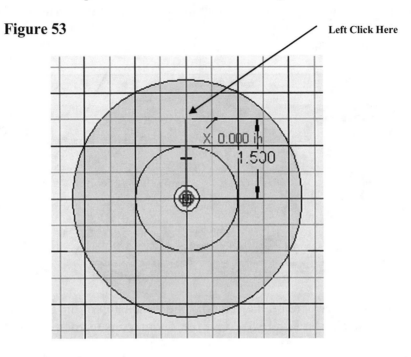

58. Move the cursor out to create a circle as shown in Figure 54.

Figure 54

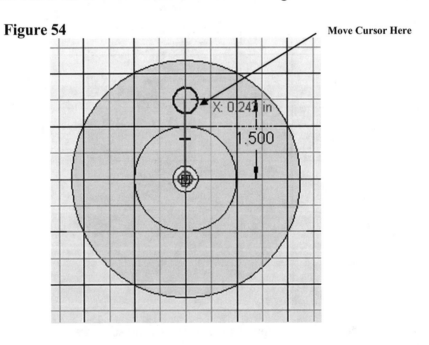

59. Left click as shown in Figure 55.

Figure 55

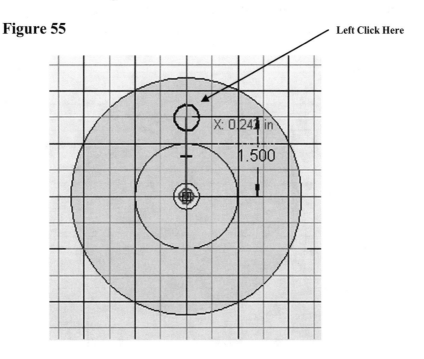

Left Click Here

60. Move the cursor to the middle left portion of the screen and left click on **SmartDimension** as shown in Figure 56.

Figure 56

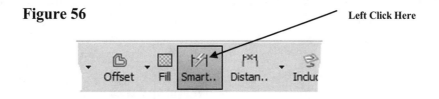

Left Click Here

61. Move the cursor to the circle that was just drawn. The circle will turn red. Select the circle by left clicking anywhere on the circle (not the center) as shown in Figure 57. The dimension box will be attached to the cursor.

Figure 57

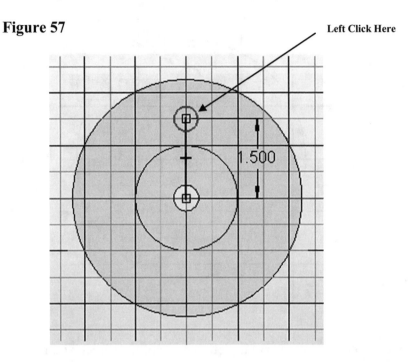

62. Move the cursor to where the dimension will be placed and left click once as shown in Figure 58.

Figure 58

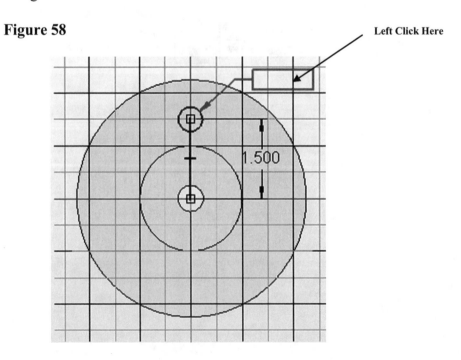

63. To edit the dimension, type **0.500** in the Dimension dialog box (while the current dimension is highlighted) as shown in Figure 59. Press **Enter** on the keyboard.

Figure 59

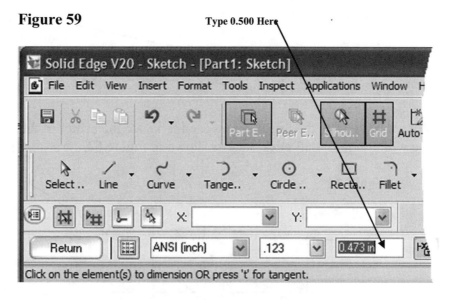

Type 0.500 Here

64. The dimension of the circle will become **0.500** inches as shown in Figure 60. Use the Zoom icons to zoom out if necessary.

Figure 60

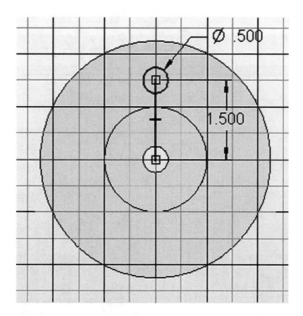

65. Move the cursor to the upper left portion of the screen and left click on **Select** as shown in Figure 61.

Figure 61

Left Click Here

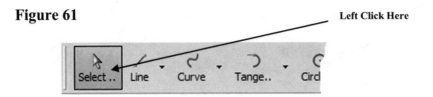

66. Move the cursor to the line that was used to locate the center of the circle. The line will turn red. Left click once as shown in Figure 62.

Figure 62

Left Click Here

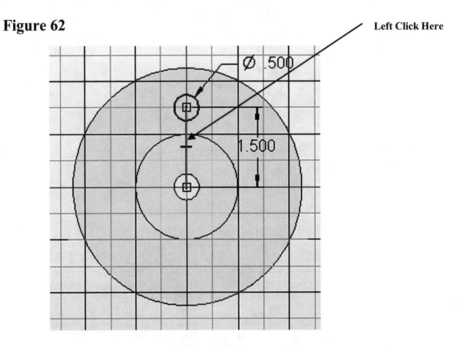

67. Move the cursor to the upper left portion of the screen and left click on **Edit**. A drop down menu will appear. Left click on **Delete** as shown in Figure 63.

Figure 63

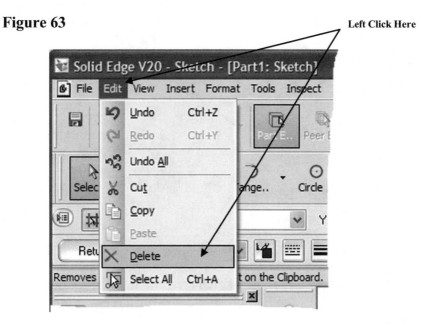

68. Solid Edge will delete the line. Your screen should look similar to Figure 64.

Figure 64

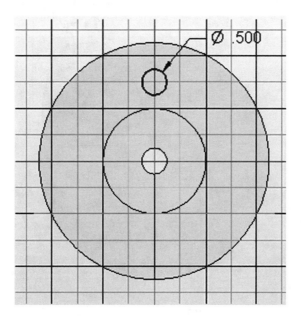

69. Move the cursor to the upper left portion of the screen and left click on **Return** as shown in Figure 65.

Figure 65

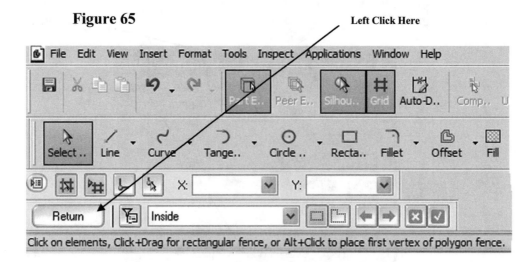

Left Click Here

70. Solid Edge is now out of the Sketch area and into the Model area. Notice that the commands at the top of the screen are now different. Your screen should look similar to Figure 66.

Figure 66

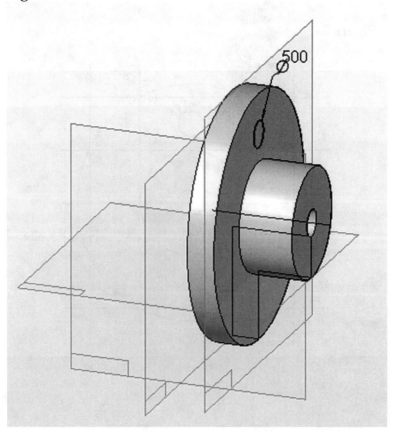

71. Move the cursor to the upper middle portion of the screen and left click on **Cutout** as shown in Figure 67.

Figure 67

Left Click Here

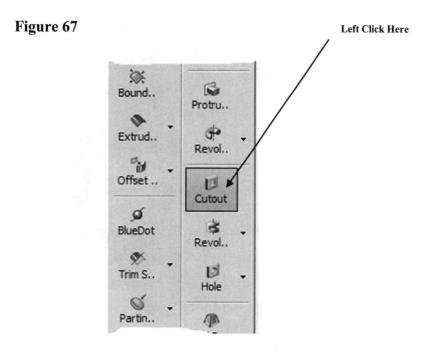

72. Move the cursor to the edge of the circle. The edge will turn red as shown in Figure 68.

Figure 68

Turned Red

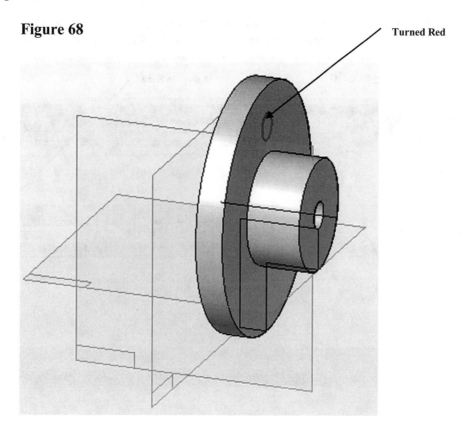

73. After the hole turns red, left click once. Then right click once. The cutout of the hole will be attached to the cursor as shown in Figure 69.

Figure 69

Hole is Attached to Cursor

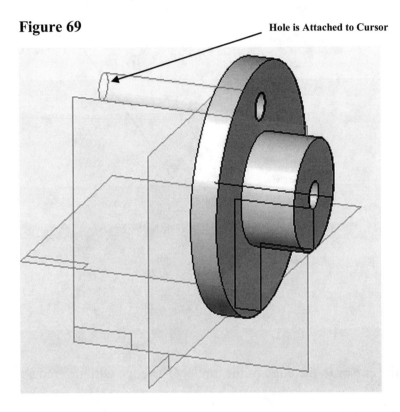

74. Move the cursor to the upper middle portion of the screen and type in **.75** in the Distance box as shown in Figure 70. Press the **Enter** key on the keyboard.

Figure 70

Enter .75 Here

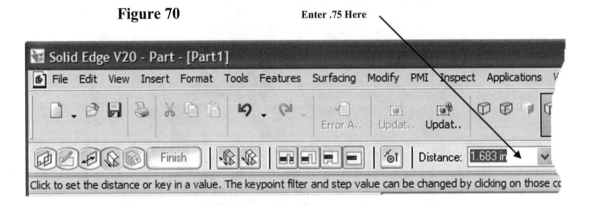

75. Notice that the depth of the hole has changed in the preview. The arrow illustrating the hole depth is now longer. Left click once as shown in Figure 71.

Figure 71   Left Click Here

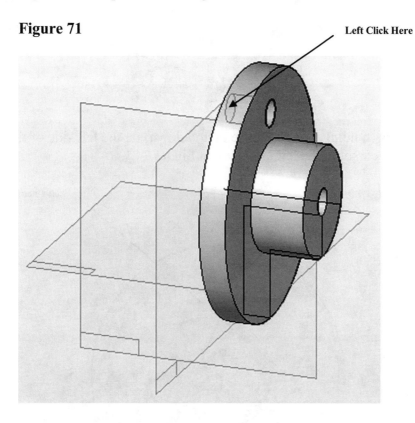

76. Move the cursor to the upper left portion of the screen and left click on **Finish** as shown in Figure 72.

Figure 72   Left Click Here

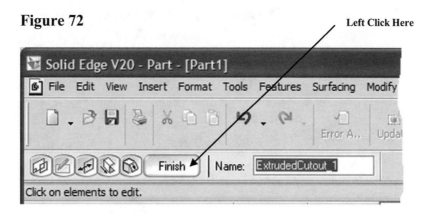

125

77. Move the cursor to the upper right portion of the screen and left click on **Rotate** as shown in Figure 73.

Figure 73

Left Click Here

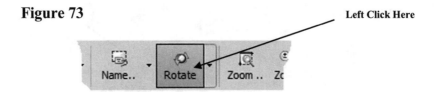

78. Left click (holding the left mouse button down) on the left side of the part and move the cursor to the left as shown in Figure 74.

Figure 74

Left Click Here

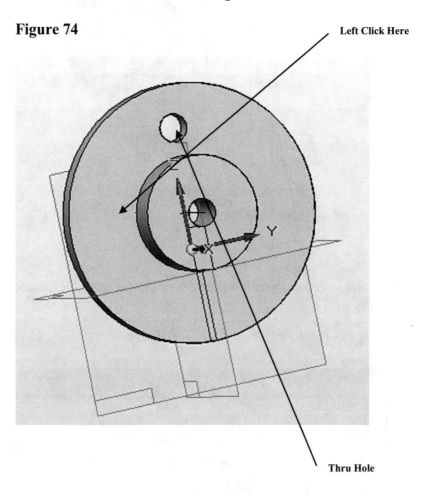

Thru Hole

79.   Your screen should look similar to Figure 74.

80.   Move the cursor to the lower middle portion of the screen and left click on the drop down arrow next to Pattern. A fly out menu will appear. Left click on **Pattern Along Curve** as shown in Figure 75.

Figure 75

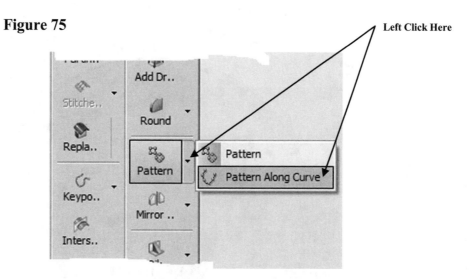

81.   Move the cursor over the hole causing it to turn red and left click once as shown in Figure 76.

Figure 76

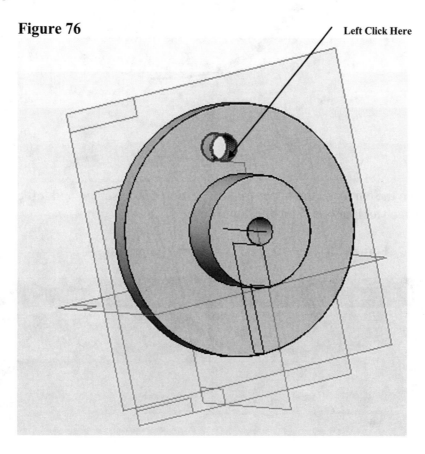

82. The hole will turn yellow as shown in Figure 77.

Figure 77

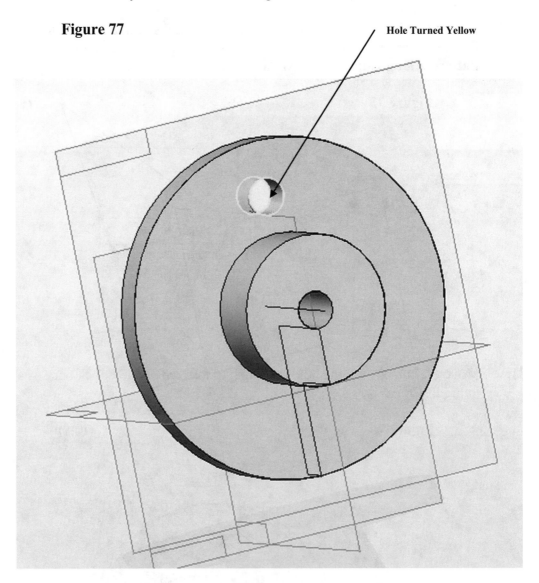

83. Move the cursor to the upper middle portion of the screen and left click on the green checkmark as shown in Figure 78.

Figure 78

84. Move the cursor to the upper middle portion of the screen and enter **Fit**, **6** for the Count and **Single** as shown in Figure 79.

Figure 79

Fit   6   Single

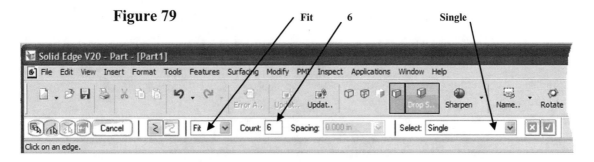

85. Move the cursor over the outside edge of the part and left click as shown in Figure 80.

Figure 80

Left Click Here

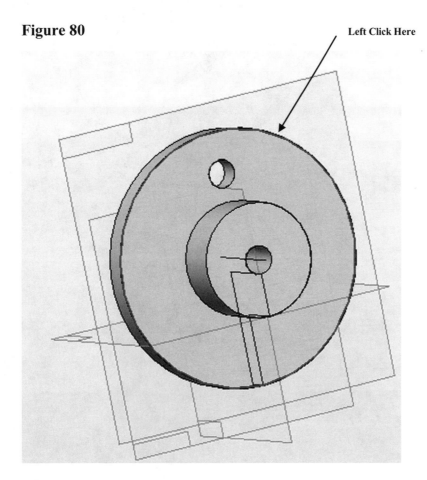

86.   An arrow will appear on the outside edge of the part. Left click as shown in Figure 81.

Figure 81

Left Click Here

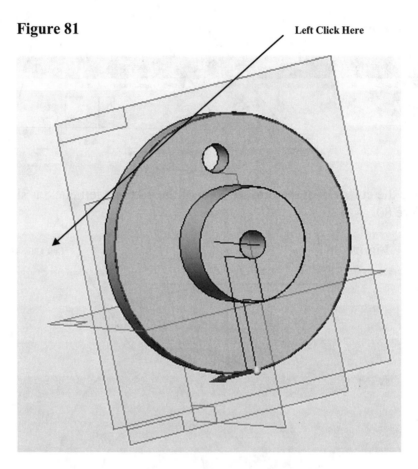

87. Solid Edge will provide a preview of the holes that will be created as shown in Figure 82.

Figure 82

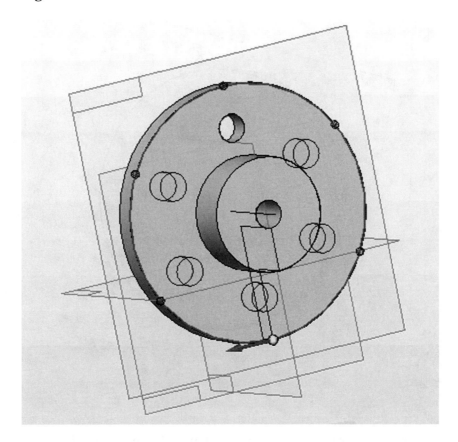

88. Move the cursor to the upper left portion of the screen and left click on **Preview** as shown in Figure 83.

Figure 83   Left Click Here

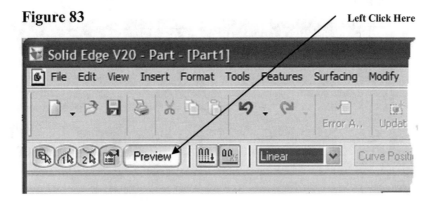

89. Solid Edge will provide a preview of the thru holes as shown in Figure 84.

Figure 84

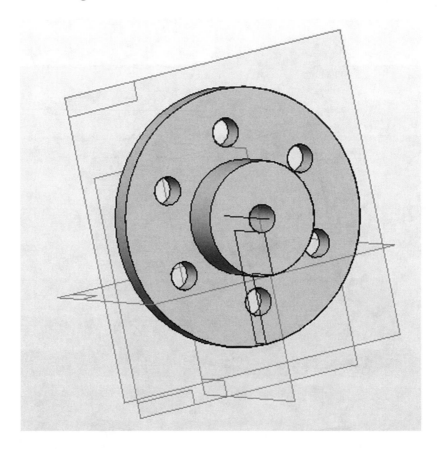

90. Move the cursor to the upper left portion of the screen and left click on **Finish** as shown in Figure 85.

Figure 85

Left Click Here

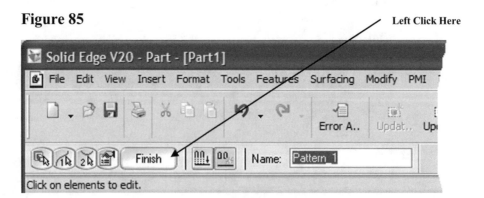

91.   Your screen should look similar to Figure 86.

Figure 86

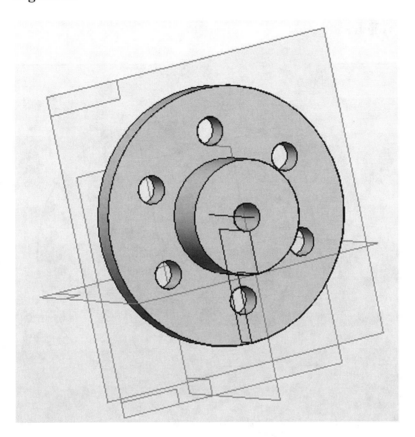

92.   To ensure that the holes are correct, move the cursor to the upper right portion of the screen and left click on the "Rotate" icon as shown in Figure 87.

Figure 87   Left Click Here

93.   The "Rotate" command will become active. Left click anywhere on the screen. Hold the left mouse button down and drag the cursor around. The part will rotate as shown in Figure 88.

Figure 88

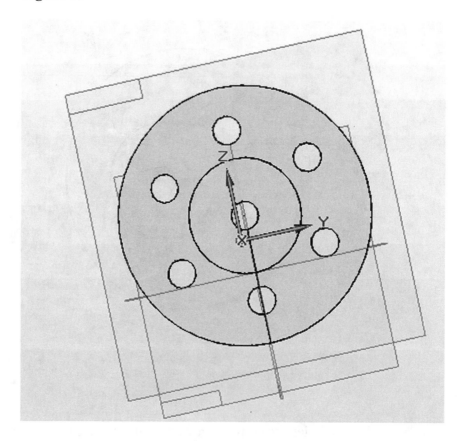

94.   Other commands for viewing are located at the top of the screen as shown in Figure 89.

Figure 89

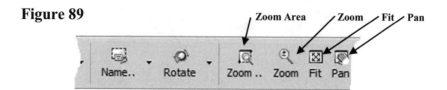

95.   The Zoom Area command works by using the cursor to draw a window around the area you want to zoom in on. After selecting the "Zoom Area" icon, hold the left mouse button down and drag a diagonal box around the desired area. Release the left mouse button when the proper amount of zoom is achieved.

96.   The Zoom command works similar to the Zoom Area command. Start by selecting the "Zoom" icon. Left click on the drawing and hold the left mouse button down while dragging the cursor up and down until the proper amount of zoom is achieved.

97.   The Pan command works similar to the Zoom command. Start by selecting the "Pan" icon. Left click on the drawing and hold the left mouse button down while moving the cursor up and down or side to side. Release the mouse button after the desired view is achieved.

98.   The Fit command works by filling the screen so that the entire part is as large as possible while remaining visible in its entirety.

99.   The same viewing options can be accessed by right clicking anywhere on the drawing. A pop up menu will be displayed as shown in Figure 90.

Figure 90

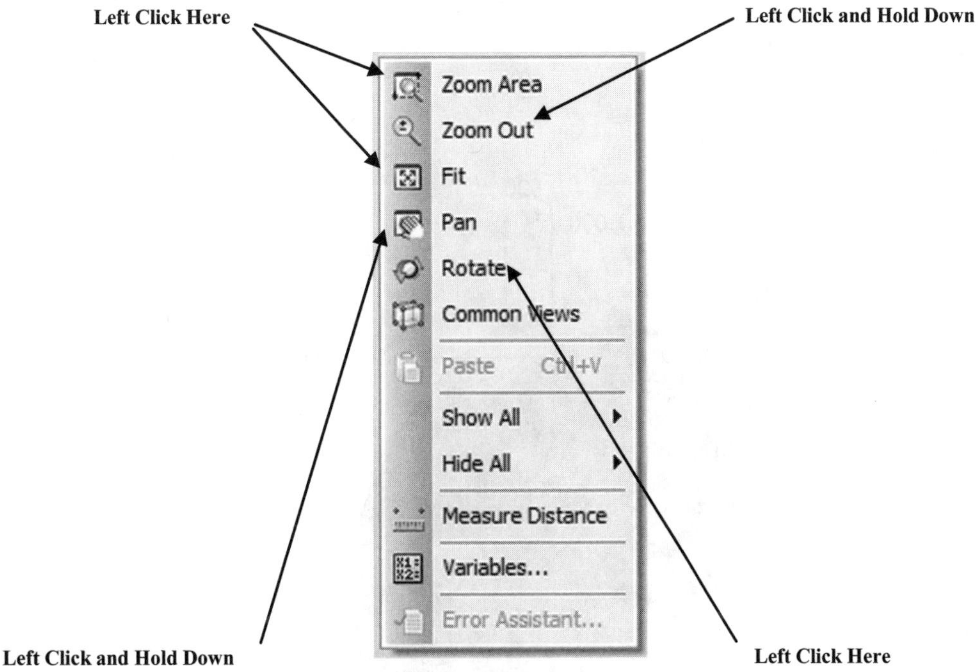

Left Click Here

Left Click and Hold Down

Zoom Area
Zoom Out
Fit
Pan
Rotate
Common Views
Paste   Ctrl+V
Show All   ▶
Hide All   ▶
Measure Distance
Variables...
Error Assistant...

Left Click and Hold Down

Left Click Here

100.   To save a file, Solid Edge must be in the Model area. If Solid Edge is in the Sketch area, the "Close" option will be inactive. To move to the Model area, move the cursor to the upper left portion of the screen and left click on **Return** as described in instruction number 69.

Drawing Activities

Problem 1

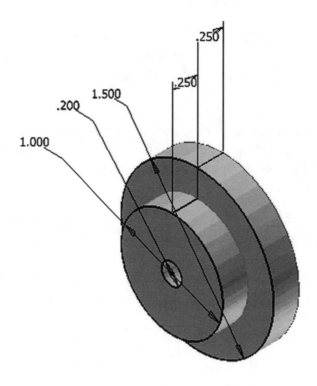

Problem 2

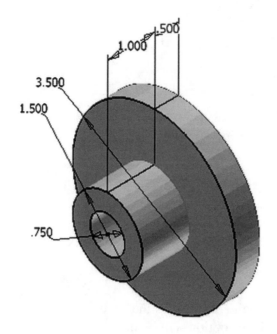

Problem 3

Revolve Axis

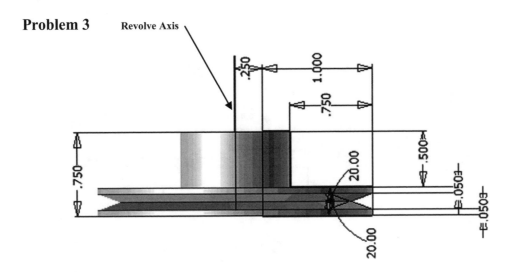

Problem 4

Revolve Axis

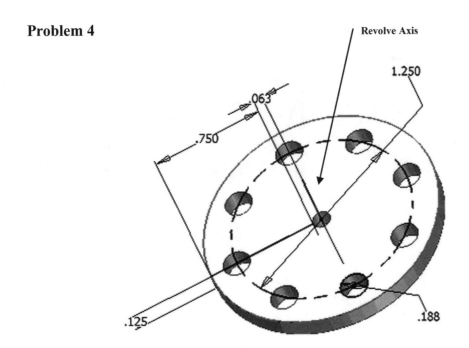

1.250

.063

.750

.125

.188

Problem 5

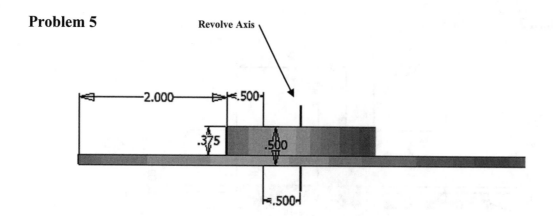

Problem 6

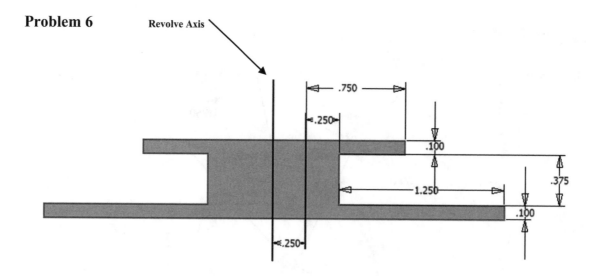

Problem 7

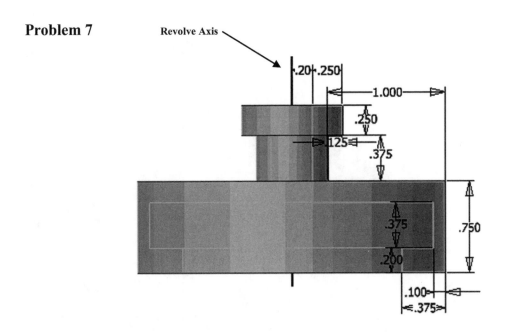

Revolve Axis

.20 .250

1.000

.250

.125

.375

.375

.750

.200

.100

.375

Problem 8

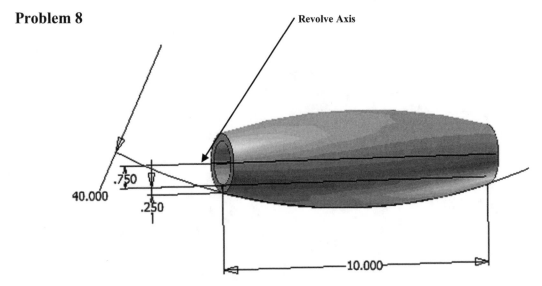

Revolve Axis

.750

40.000

.250

10.000

Chapter 3 Learning To Create a Detail Drawing

Objectives:

- Create a simple sketch using the Sketch command
- Extrude a sketch into a solid using the Model command
- Create an Orthographic view using the Drawing View Wizard
- Create an Isometric View

Chapter 3 includes instruction on how to design the parts shown below.

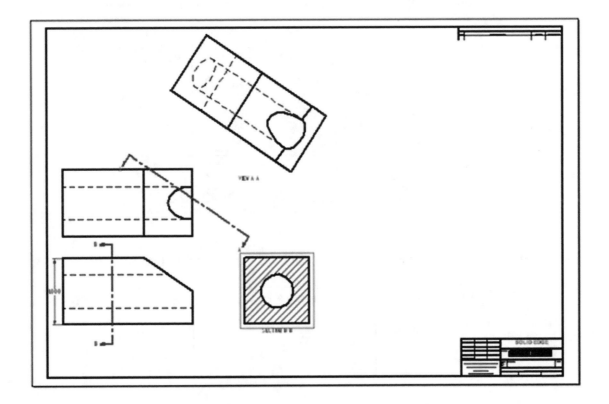

1.	Start Solid Edge by referring to "Chapter 1 Getting Started".

2.	After Solid Edge is running, begin a new sketch.

3.	Move the cursor to the upper left corner of the screen and left click on **Line** as shown in Figure 1.

Figure 1

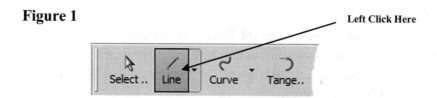

Left Click Here

4.	Move the cursor somewhere in the lower left portion of the screen and left click once. This will be the beginning end point of a line as shown in Figure 2.

Figure 2

Left Click Here

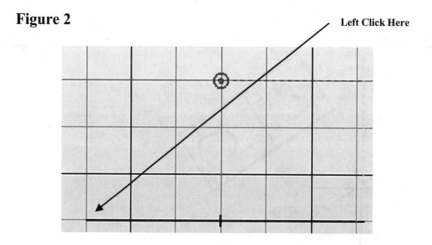

5.	Move the cursor to the right and left click once as shown in Figure 3.

Figure 3

Left Click Here

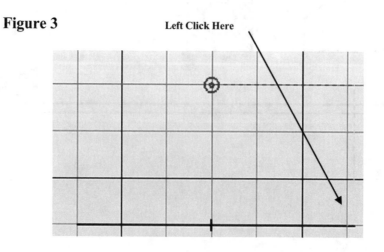

6.   Move the cursor upward and left click once as shown in Figure 4.

Figure 4

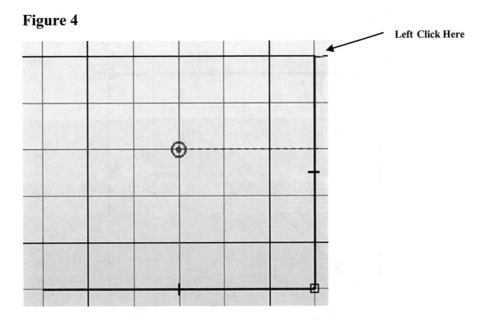

Left Click Here

7.   Move the cursor to the left and wait for dashes to appear. Left click once as shown in Figure 5.

Figure 5

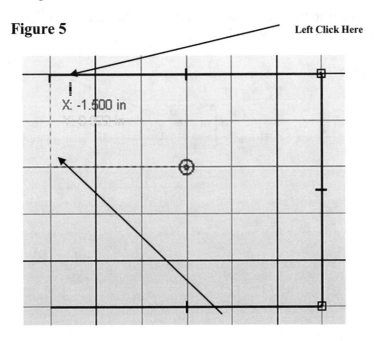

Left Click Here

X: -1.500 in

Dashes Signify End Points are Aligned

8.   Move the cursor back to the original starting end point. A red dot will appear.
      Left click once. Your screen should look similar to Figure 6.

Figure 6

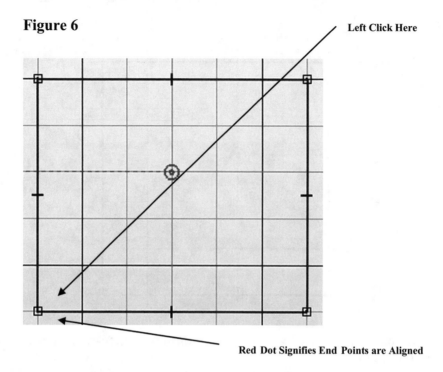

Left Click Here

Red Dot Signifies End Points are Aligned

9.   Move the cursor to the upper middle portion of the screen and left click
      on **SmartDimension** as shown in Figure 7.

Figure 7

Left Click Here

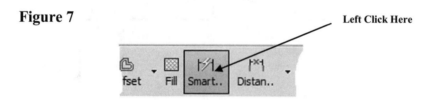

10. After selecting **SmartDimension** move the cursor over the bottom horizontal line until it turns red as shown in Figure 8. Select the line by left clicking anywhere on the line **or** on each of the end points. The dimension box will be attached to the cursor.

Figure 8   Turned Red

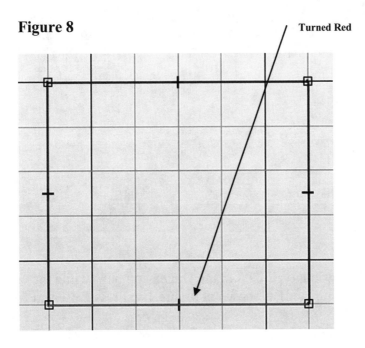

11. Move the cursor to where the dimension will be placed and left click once as shown in Figure 9.

Figure 9   Left Click Here

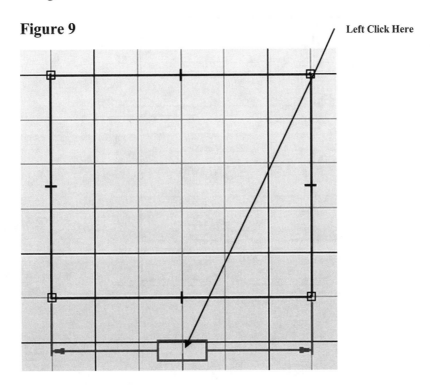

12.   The Dimension dialog box will appear as shown in Figure 10.

Figure 10

Enter 2.00 Here

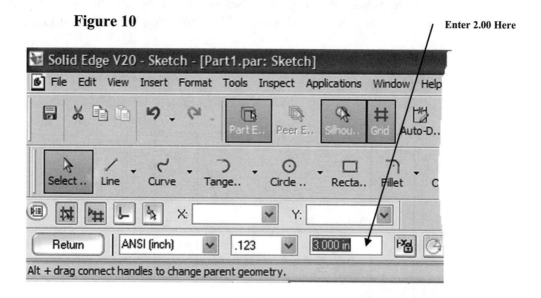

13.   To edit the dimension, type **2.00** in the Dimension box (while the current dimension is highlighted) as shown in Figure 10. Press **Enter** on the keyboard.

14.   The dimension of the line will become 2.00 inches as shown in Figure 11.

Figure 11

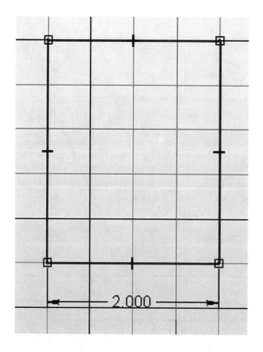

15. To view the entire drawing, move the cursor to the upper right portion of the screen and left click once on the "Fit" icon as shown in Figure 12.

Figure 12

Left Click Here

16. The drawing will "fill up" the entire screen. If the drawing is still too large, left click on the "Zoom" icon as shown in Figure 13. Hold the left mouse button down and move the cursor up and down to achieve the desired view of the sketch.

Figure 13

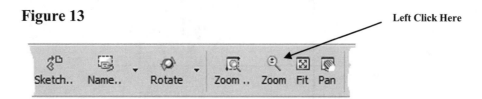

Left Click Here

17. Move the cursor to the upper left portion of the screen and left click on **SmartDimension** as shown in Figure 14.

Figure 14

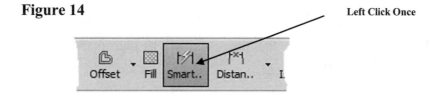

Left Click Once

18. Move the cursor over the right side line until it turns red as shown in Figure 15. Select the line by left clicking anywhere on the line **or** on each of the end points. The dimension box will be attached to the cursor.

Figure 15

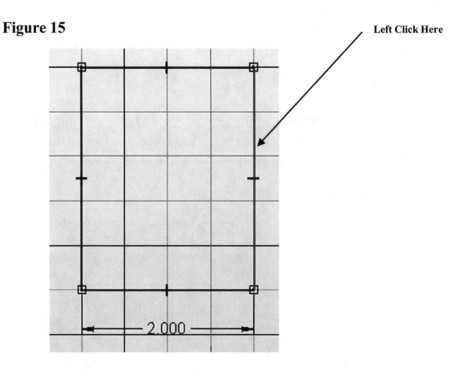

Left Click Here

2.000

19. Move the cursor to where the dimension will be placed and left click once as shown in Figure 16.

Figure 16   Left Click Here

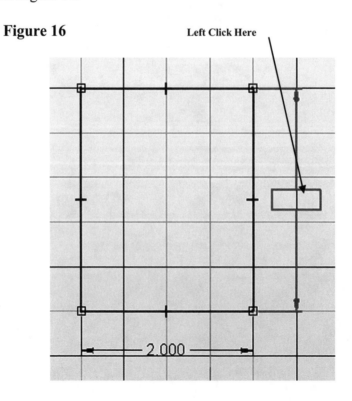

2.000

20. While the dimension is highlighted type **1.0** in the dimension box as shown in Figure 17. Press the **Enter** key on the keyboard. Press the **Esc** key once or twice to exit the SmartDimension command.

Figure 17

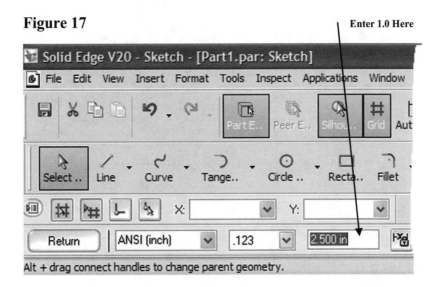

21. The dimension of the line will become 1.00 inches as shown in Figure 18. Use the Zoom icons to zoom out if necessary.

Figure 18

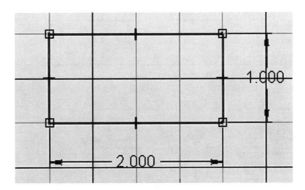

22. Move the cursor over the top horizontal line until it turns red as shown in Figure 19. Select the line by left clicking anywhere on the line **or** on each of the end points. This will cause the dimension box to be attached to the cursor.

Figure 19

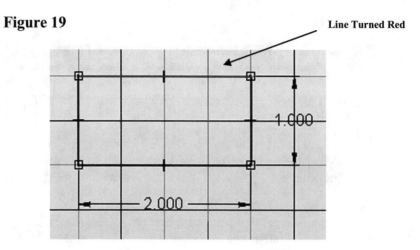

23. Move the cursor to where the dimension will be placed and left click once as shown in Figure 20.

Figure 20

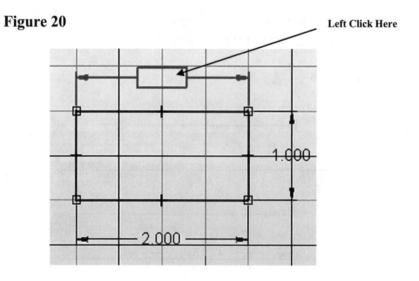

24.   Notice that the dimension is exactly 2.000 as shown in Figure 21.

Figure 21

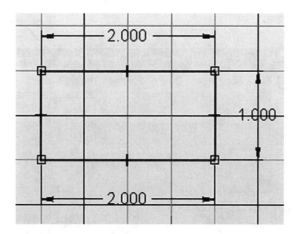

25.   The dimension will appear in blue indicating that this length has already been dimensioned as shown by the bottom dimension in Figure 22. It is not necessary to dimension this length again.

Figure 22   Blued Dimension

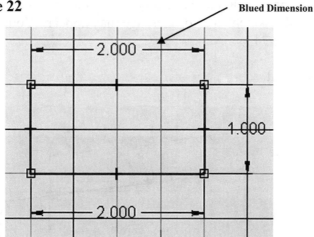

26.   Dimensioning the far left line would also result in a blued dimension. Because of this, the dimensioning portion is complete.

27.   After the sketch is complete it is time to extrude the sketch into a solid.

28. After you have verified that no commands are active, move the cursor to the upper left portion of the screen and left click on **Return** as shown in Figure 23.

Figure 23

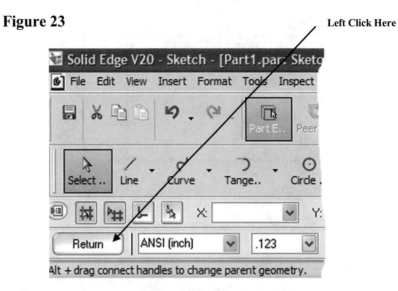

Left Click Here

29. Solid Edge is now out of the Sketch area and into the Model area. Notice that the commands at the top of the screen are now different. To work in the Model area a sketch must be present and have no opens (non-connected lines). If there are any opens in the sketch an error message may appear. Your screen should look similar to Figure 24.

Figure 24

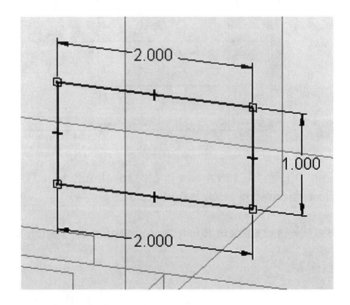

30.   Move the cursor to the upper left portion of the screen and left click on
        Protrusion as shown in Figure 25. If Solid Edge gave you an error message,
        there are opens (non-connected lines) somewhere on the sketch. Check each
        intersection for opens by using the **Extend** and **Trim** commands.

Figure 25

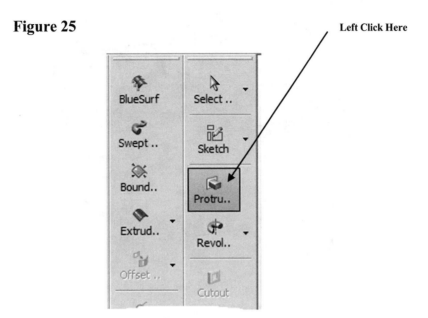

31.   Move the cursor over the top line causing the entire box to turn red. Left click
        then right click once as shown in Figure 26.

Figure 26

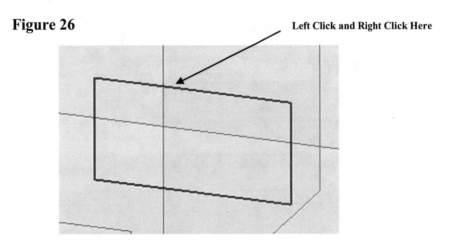

32. The sketch will become three dimensional as shown in Figure 27. The extruded surface will be attached to the cursor.

Figure 27

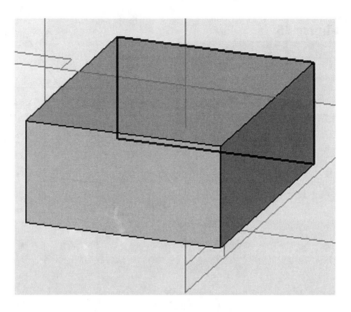

33. Move the cursor to the upper middle portion of the screen and type **1.00** next to the text "Distance" as shown in Figure 28.

Figure 28

Type 1.00 Here

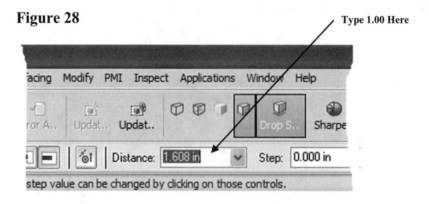

34. Left click anywhere around the sketch. Solid Edge will create a solid from the sketch as shown in Figure 29.

Figure 29

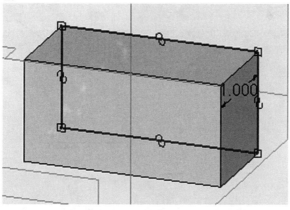

35. Move the cursor to the upper left portion of the screen and left click on **Finish** as shown in Figure 30.

Figure 30

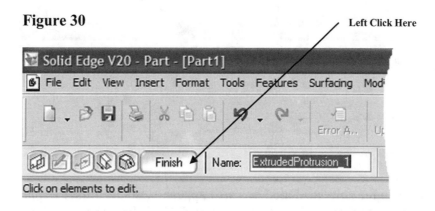

36. The screen should look similar to Figure 31.

Figure 31

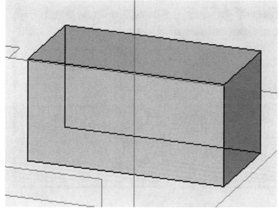

37. Move the cursor to the middle left portion of the screen and left click on the drop down arrow to the right of the "Round" icon. A fly out menu will appear. Left click on **Chamfer** as shown in Figure 32.

Figure 32

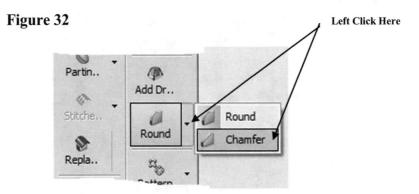

38. Move the cursor to the upper left portion of the screen and left click on the "Chamfer Options" icon as shown in Figure 33.

Figure 33

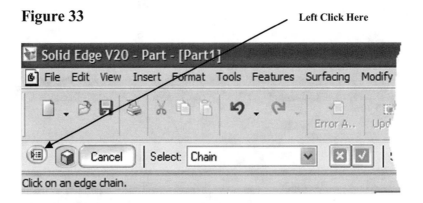

39. The Chamfer Options dialog box will appear. Use the cursor to place a dot next to the text "2 Setbacks" and left click on **OK** as shown in Figure 34.

Figure 34

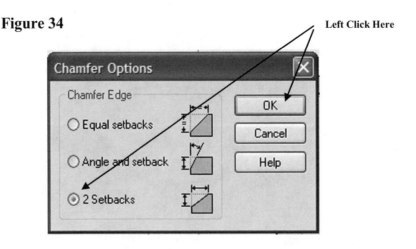

40. Move the cursor to the upper left portion of the screen and left click on the drop down arrow to the right of the text "Select". A drop down menu will appear. Left click on **Face** as shown in Figure 35.

Figure 35

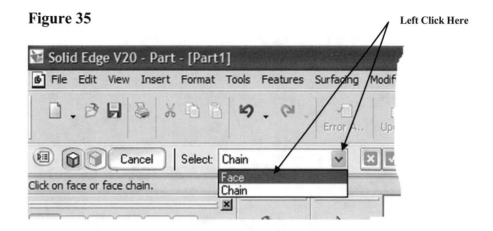

41. Move the cursor over the face causing it to turn red. Left click once then right click as shown in Figure 36.

Figure 36

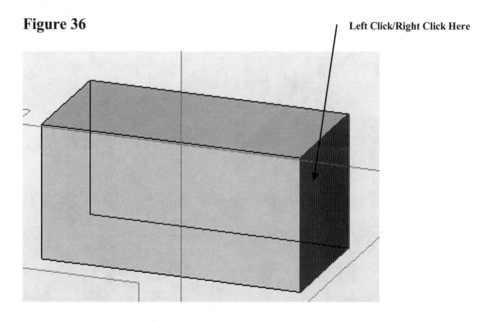

42. Move the cursor to the upper left portion of the screen and left click on the drop down arrow to the right of the text "Select". A drop down menu will appear. Left click **Edge/Corner** as shown in Figure 37.

Figure 37

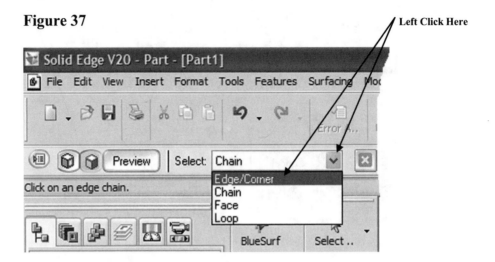

43. Move the cursor to the edge shown in Figure 38 causing it to turn red. Left click once.

Figure 38

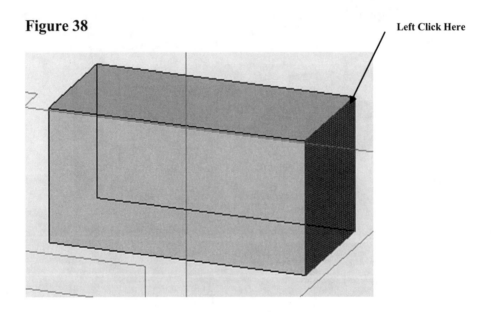

44.   Move the cursor to the upper middle portion of the screen to the right of the text "Setbacks: 1:" and enter **.500** and **.750** as shown in Figure 39. Right click once.

Figure 39

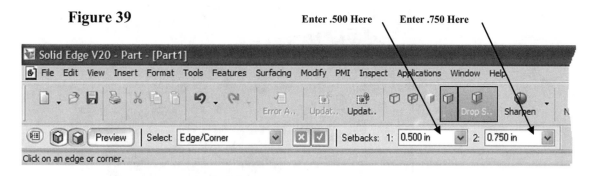

45.   A preview of the chamfer will be displayed as shown in Figure 40.

Figure 40   **Anticipated Chamfer**

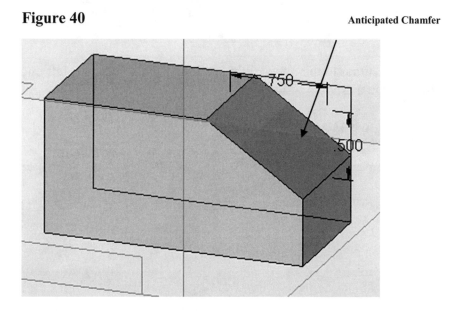

46.   Move the cursor to the upper left portion of the screen and left click on **Finish** as shown in Figure 41.

Figure 41   **Left Click Here**

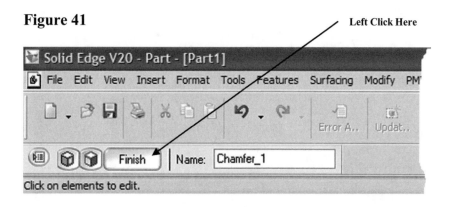

47. Your screen should look similar to Figure 42.

Figure 42

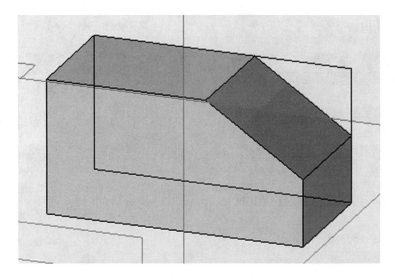

48. Move the cursor to the upper right portion of the screen and left click on the **Rotate** icon as shown in Figure 43.

Figure 43   Left Click Here

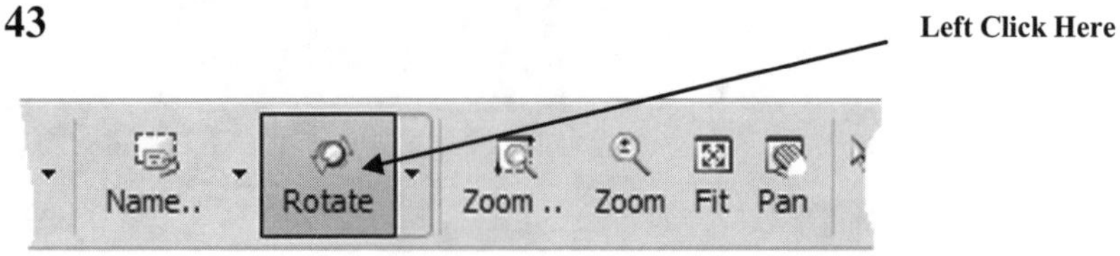

49. While holding the left mouse button down, drag the cursor to the right to gain access to the backside of the part as shown in Figure 44.

Figure 44

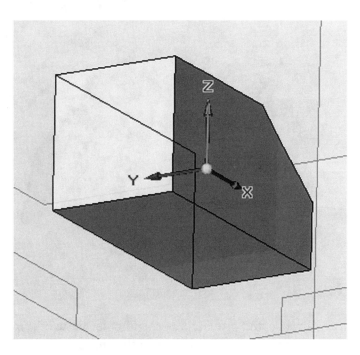

50. Right click anywhere on the screen. Right clicking is the same as pressing the **Esc** key on the keyboard.

51. Move the cursor to the upper left portion of the screen and left click on **Sketch** as shown in Figure 45.

Figure 45

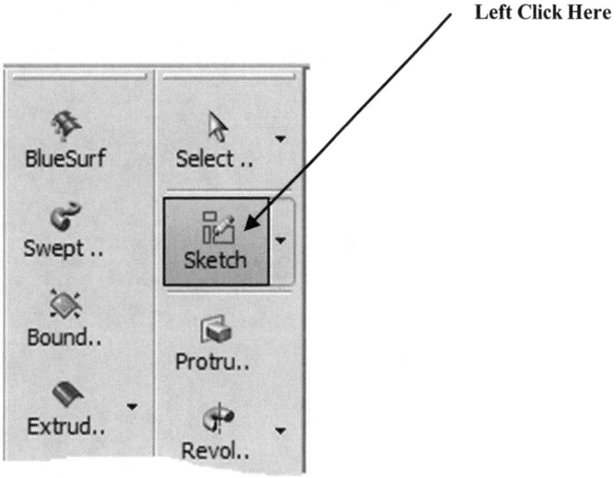

Left Click Here

BlueSurf   Select ..

Swept ..   Sketch

Bound..   Protru..

Extrud..   Revol..

52. Move the cursor to the face shown in Figure 46 causing it to turn red. Left click once.

Figure 46

Left Click Here

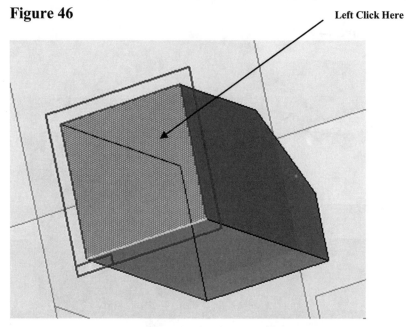

53. The part will rotate and provide a perpendicular view of the surface as shown in Figure 47. A new sketch will appear on the surface.

Figure 47

Perpendicular View

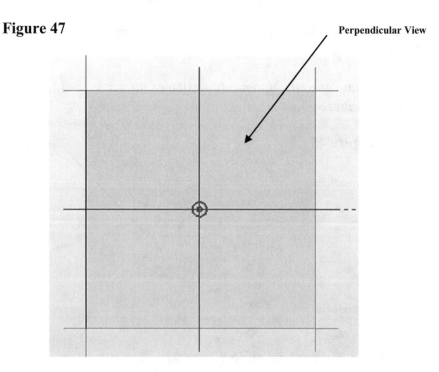

54. Move the cursor to the upper left portion of the screen and left click on **Circle** as shown in Figure 48.

Figure 48

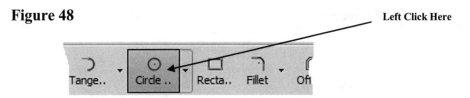

55. Left click on the center of the part on the backside surface as shown in Figure 49.

Figure 49

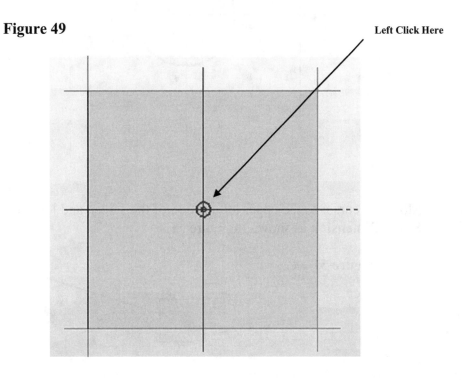

56. Move the cursor to the side of the circle and left click once as shown in Figure 50.

Figure 50

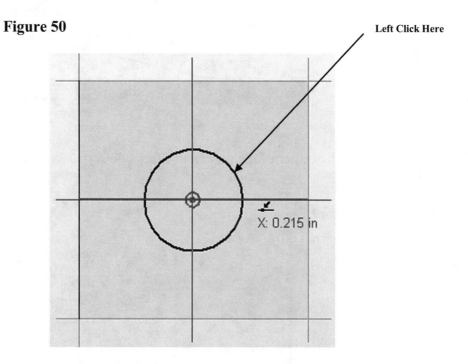

Left Click Here

X: 0.215 in

57. Move the cursor to the upper left portion of the screen and left click on **SmartDimension** as shown in Figure 51.

Figure 51

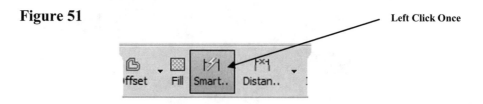

Left Click Once

ffset   Fill Smart.. Distan..

58. After selecting **SmartDimension** move the cursor over the edge of the circle until it turns red as shown in Figure 52. Left click on the edge of the circle.

Figure 52

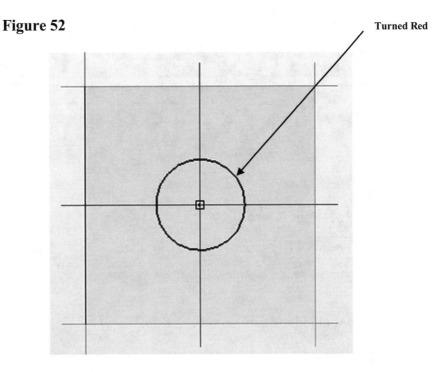

Turned Red

59. Move the cursor to where the dimension will be placed and left click once as shown in Figure 53.

Figure 53

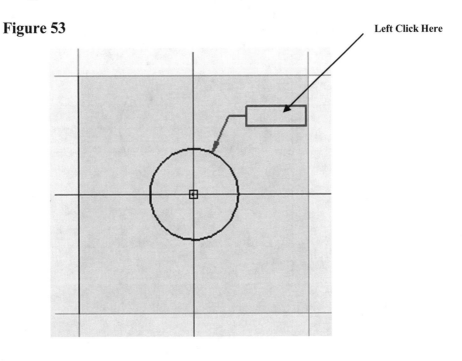

Left Click Here

60. To edit the dimension, type **.50** in the Dimension box (while the current dimension is highlighted) as shown in Figure 54. Press **Enter** on the keyboard.

Figure 54

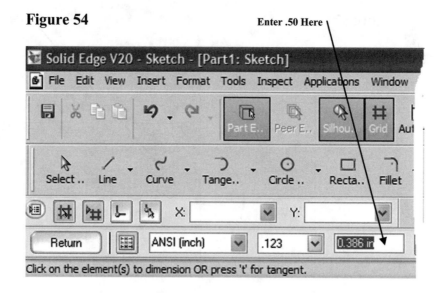

61. Your screen should look similar to Figure 55.

Figure 55

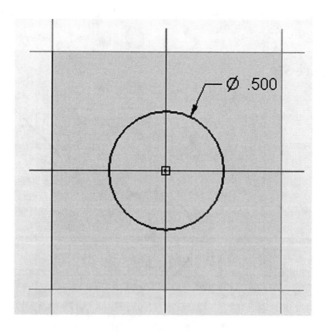

62. Move the cursor over the edge of the part until it turns red. Select the line by left clicking anywhere on the edge as shown in Figure 56.

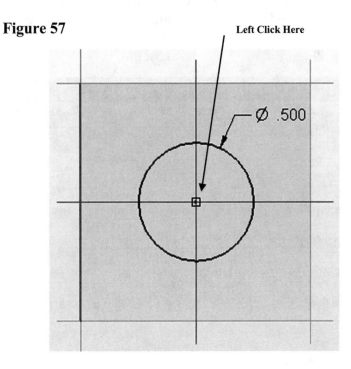

Figure 56

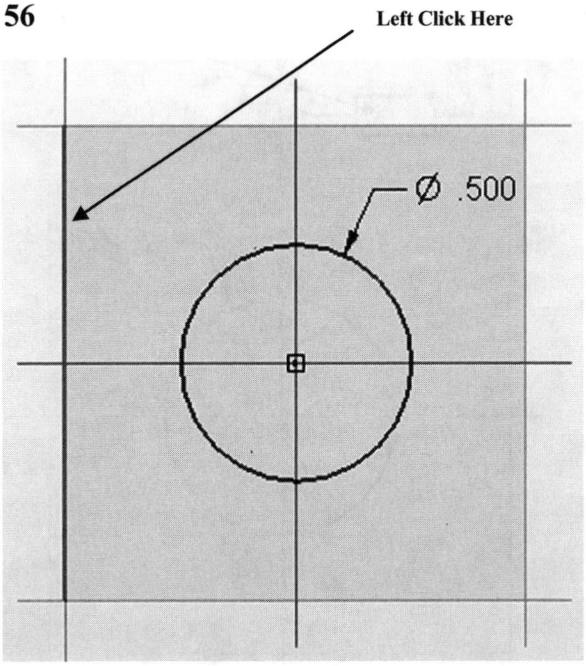

Left Click Here

Ø .500

63. Move the cursor over the center of the circle until it turns red. Left click as shown in Figure 57.

Figure 57

Left Click Here

Ø .500

64. Move the cursor to where the dimension will be placed and left click once as shown in Figure 58.

Figure 58

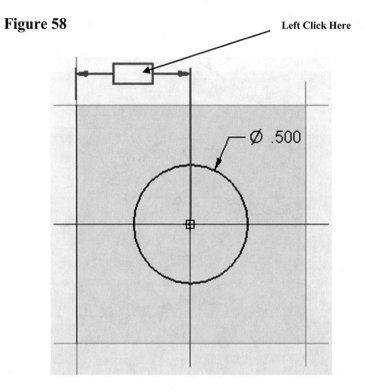

Left Click Here

Ø .500

65. To edit the dimension, type **.500** in the Dimension box (while the current dimension is highlighted) as shown in Figure 59. Press **Enter** on the keyboard.

.

Figure 59

Enter .500 Here

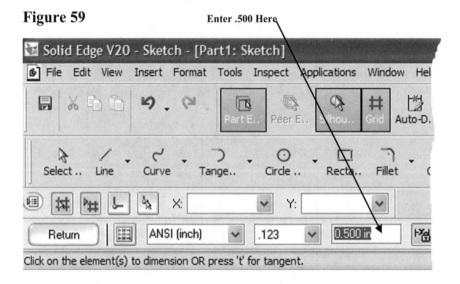

66.   Your screen should look similar to Figure 60.

Figure 60

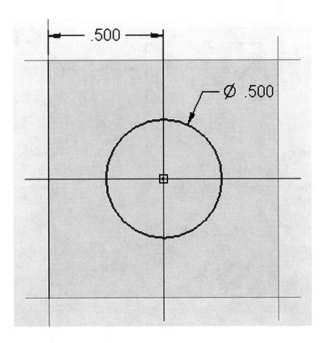

67.   Move the cursor to the edge of the part until it turns red. Left click anywhere on the edge as shown in Figure 61.

Figure 61

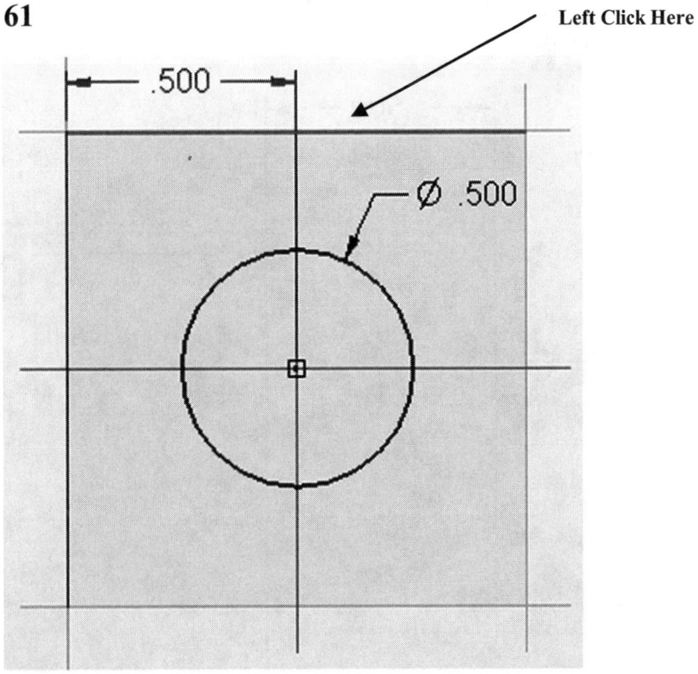

68. Move the cursor over the center of the circle until it turns red. Left click as shown in Figure 62.

Figure 62

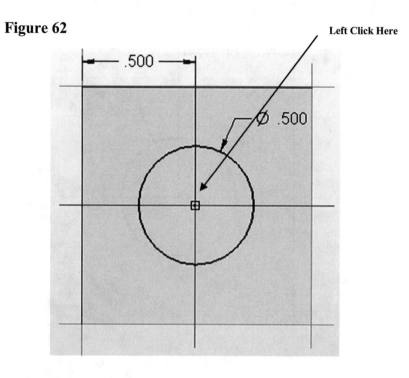

69. Move the cursor to where the dimension will be placed and left click once as shown in Figure 63.

Figure 63

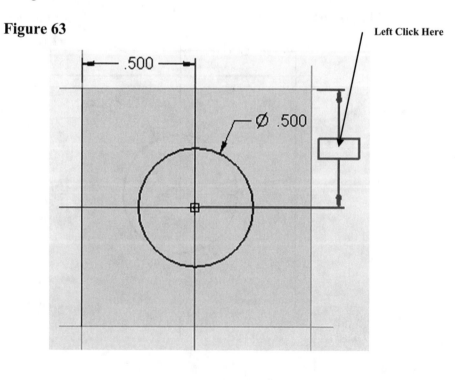

70. To edit the dimension, type **.500** in the Dimension box (while the current dimension is highlighted) as shown in Figure 64. Press **Enter** on the keyboard.

Figure 64   Left Click Here

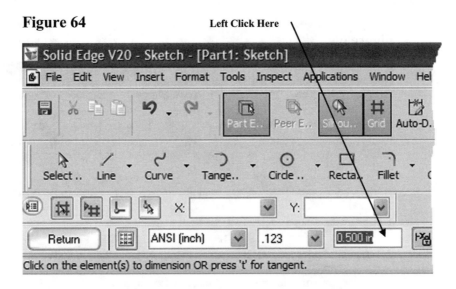

71. Your screen should look similar to Figure 65.

Figure 65

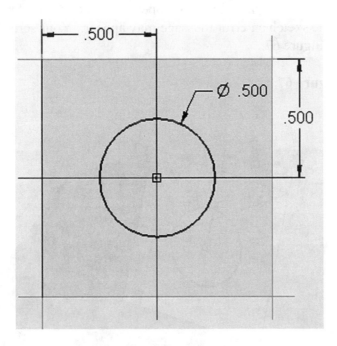

72. After the sketch is complete it is time to create a hole through the solid.

73. After you have verified that no commands are active, move the cursor to the upper left portion of the screen and left click on **Return** as shown in Figure 66.

Figure 66

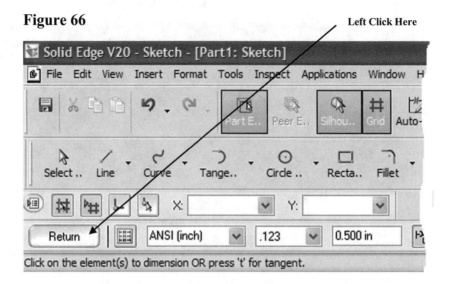

74. Solid Edge is now out of the Sketch area and into the Model area. Notice that the commands at the top of the screen are now different. To work in the Model area a sketch must be present and have no opens (non-connected lines). If there are any opens in the sketch an error message may appear. Your screen should look similar to Figure 67.

Figure 67

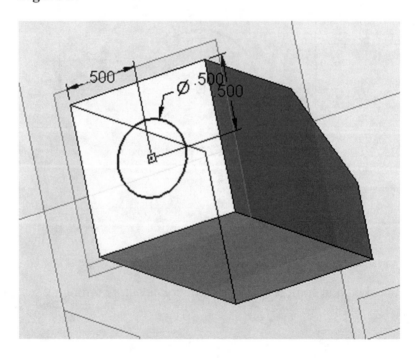

75. Move the cursor to the upper middle portion of the screen and left click on the "Rotate" icon as shown in Figure 68.

Figure 68

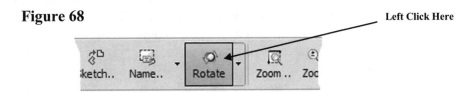

Left Click Here

76. While holding the left mouse button down, drag the cursor to the right to gain an isometric view of the part as shown in Figure 69.

Figure 69

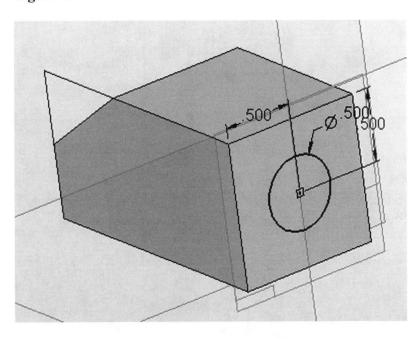

77. Move the cursor to the upper middle portion of the screen and left click on **Cutout** as shown in Figure 70. If you received an error message, there are opens (non-connected lines) somewhere on the sketch. Check each intersection for opens by using the **Extend** and **Trim** commands.

Figure 70

Left Click Here

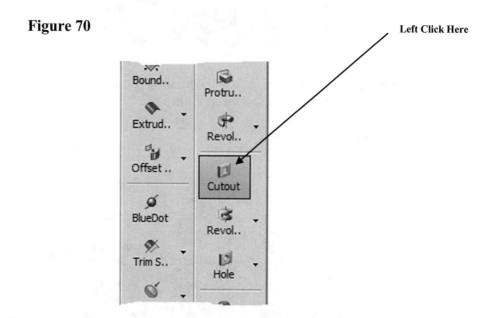

78. Move the cursor over the edge of the circle causing it to turn red. Left click on the edge of the circle, then right click as shown in Figure 71.

Figure 71

Left Click/Right Click Here

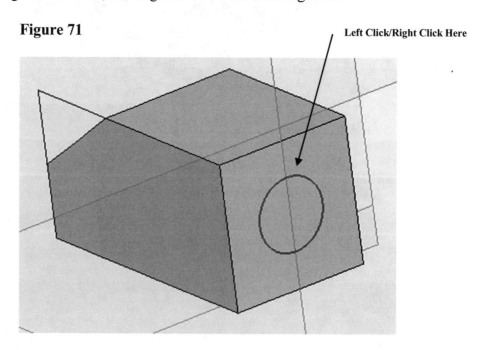

79.   A preview of the cutout will be displayed. The cutout will be attached to the cursor as shown in Figure 72.

Figure 72   **Attached to Cursor**

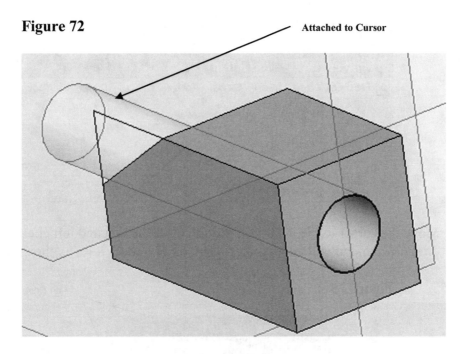

80.   To edit the dimension, type **2.00** in the Dimension box (while the current dimension is highlighted) as shown in Figure 73. Press **Enter** on the keyboard. Left click anywhere around the part.

Figure 73   **Enter 2.00 Here**

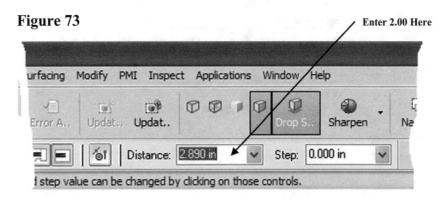

81.   Move the cursor to the upper left portion of the screen and left click on **Finish** as shown in Figure 74.

Figure 74

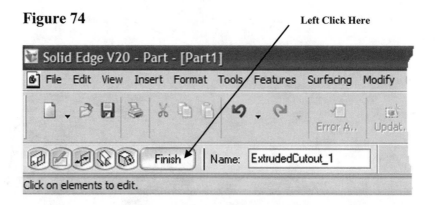

82.   Move the cursor to the upper left portion of the screen and left click on **View**. A drop down menu will appear. Left click on **Named Views** as shown in Figure 75.

Figure 75

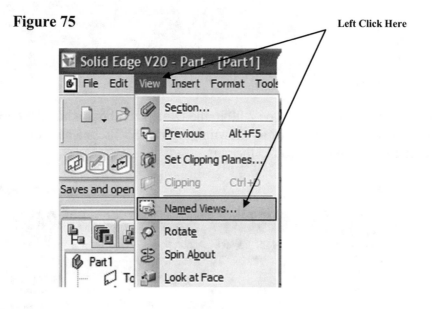

83.   The Named Views dialog box will appear. Left click on **dimetric** then left click on **Apply** as shown in Figure 76.

Figure 76

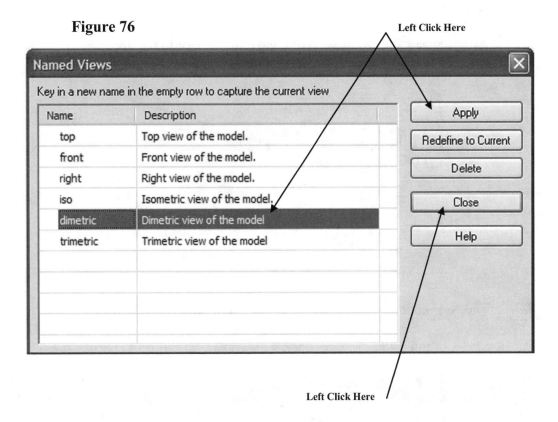

84.   Left click on **Close**.

85.   Your screen should look similar to Figure 77.

Figure 77

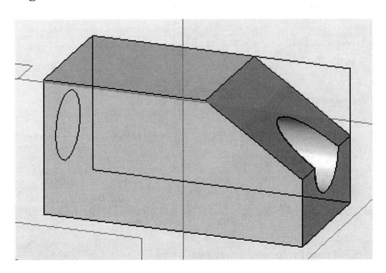

86. Save the part file as Part1.par for easy retrieval to be used in the following section.

87. After the part file has been saved, move the cursor to the upper left portion of the screen and left click on the "New" icon as shown in Figure 78.

Figure 78

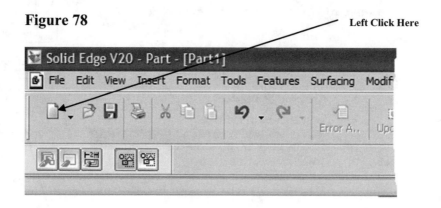

Left Click Here

88. The New dialog box will appear as shown in Figure 79.

Figure 79

Left Click Here

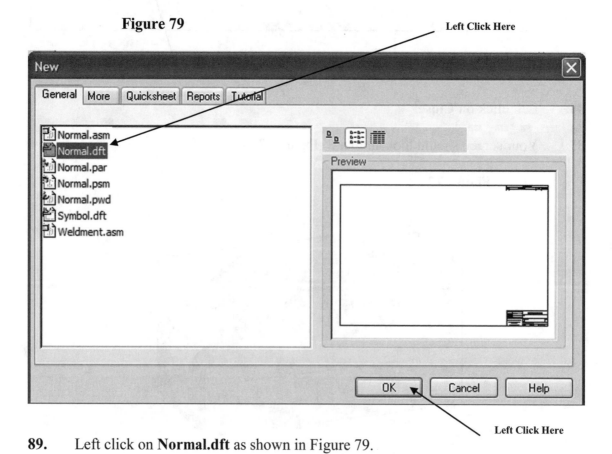

Left Click Here

89. Left click on **Normal.dft** as shown in Figure 79.

90. Left click on **OK**.

91. Move the cursor to the upper middle portion of the screen and left click on **Drawing** as shown in Figure 80.

Figure 80

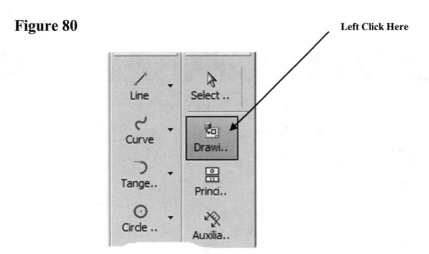

Left Click Here

92. The Select Model dialog box will appear. Left click on **Part1.par**. Left click on **Open** as shown in Figure 81.

Figure 81

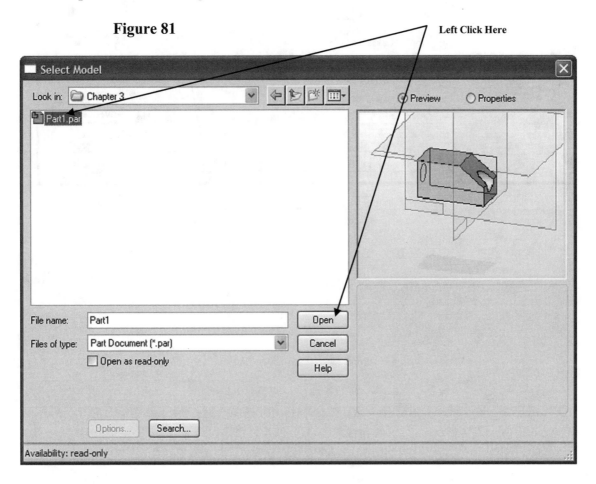

Left Click Here

93. The Drawing View Creation Wizard dialog box will appear. Left click on **Next** as shown in Figure 82.

Figure 82

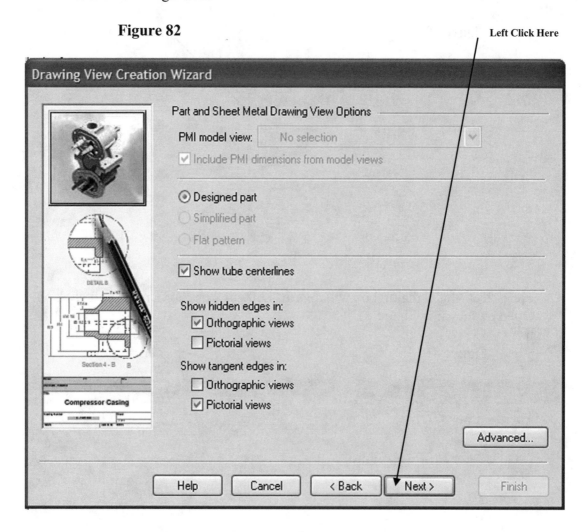

94. Left click on **front** and **Next** as shown in Figure 83.

Figure 83

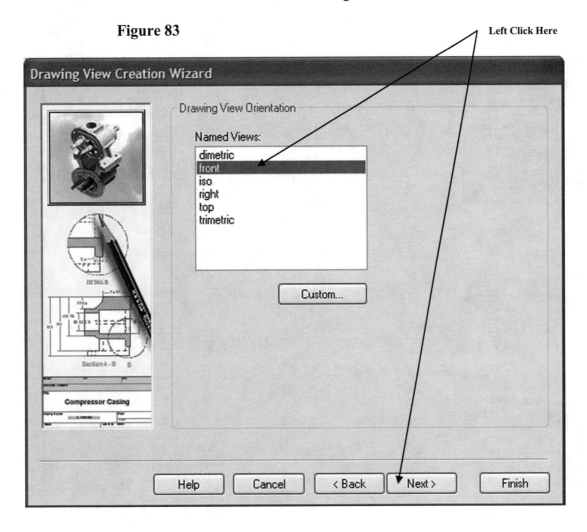

95.   Left click on the "Top", "Side" and "Isometric" view icons as shown in Figure 84.

Figure 84

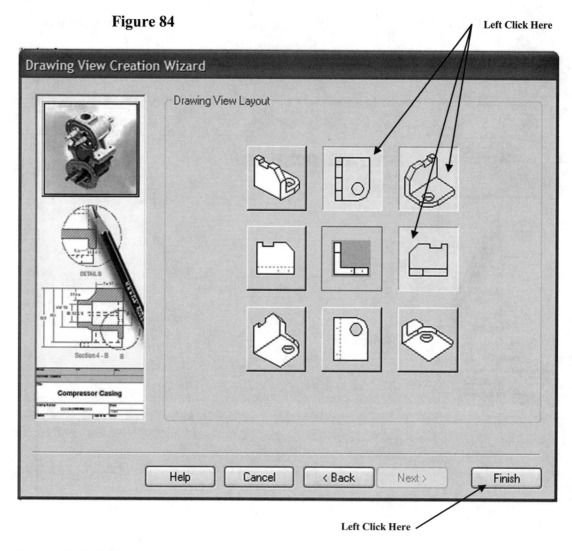

96.   Left click on **Finish**.

97.   A red box will be attached to the cursor. Move the red box to the location shown in Figure 85 and left click once.

Figure 85

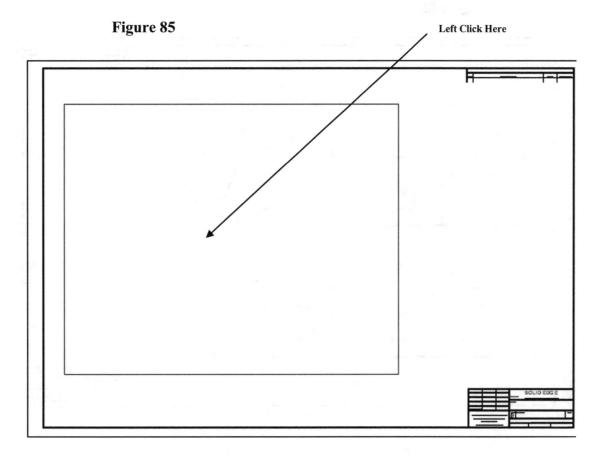

98. Your screen should look similar to Figure 86.

Figure 86

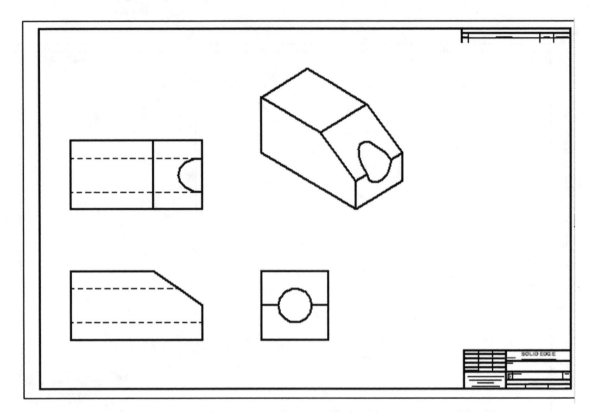

99.   If there is ever a need to move the views, simply move the cursor over one of the views causing a red box to appear around the view as shown in Figure 87. Left click (holding the left mouse button down) and drag the view to the desired location.

Figure 87   Red Box

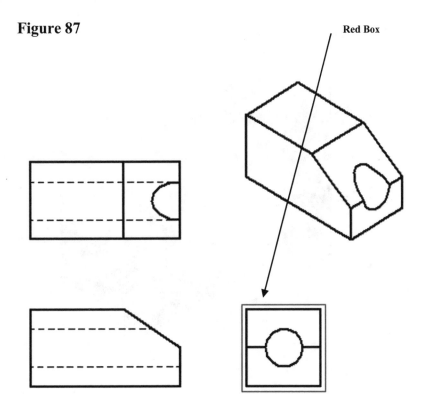

100.   Save the part file as Part1.dft for easy retrieval. This part will be used in the following chapter.

Drawing Activities

Use these problems from Chapters 1 and 2 to create three view orthographic detail drawings.

Problem 1

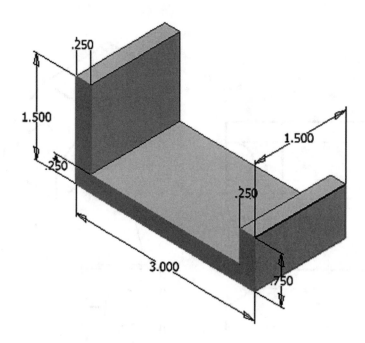

Problem 2

Extrude Center Section .25 Deep

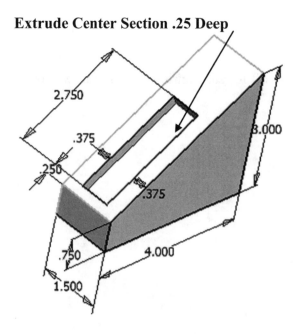

Problem 3

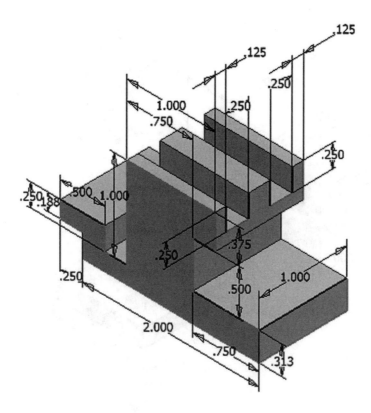

Problem 4

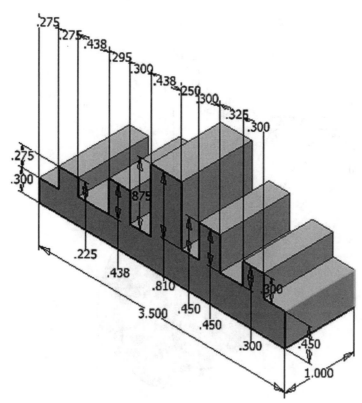

Problem 5

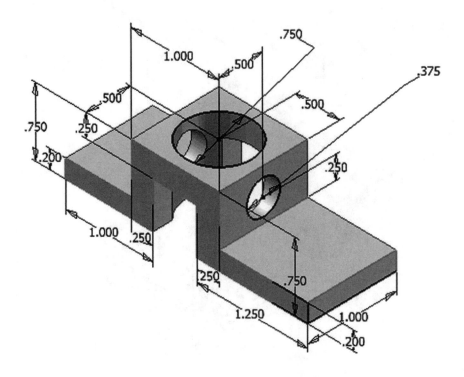

Problem 6

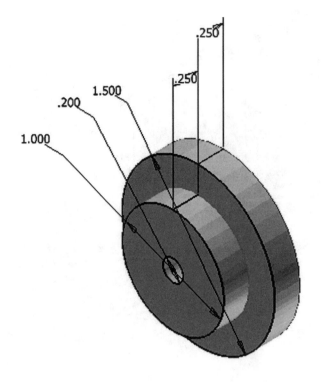

Problem 7

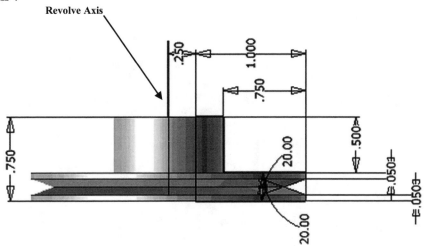

Problem 8

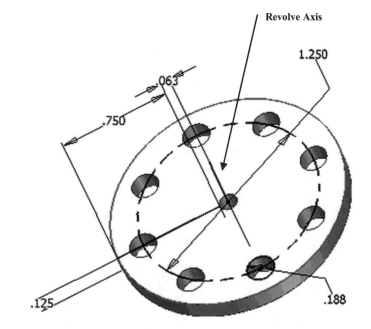

Chapter 4 Advanced Detail Drawing Procedures

Objectives:

- Create an Auxiliary View using the Auxiliary View command
- Dimension views using the SmartDimension command
- Create a Section View using the Section View command
- Create Text using the Text Profile command

Chapter 4 includes instruction on how to create the drawing shown below.

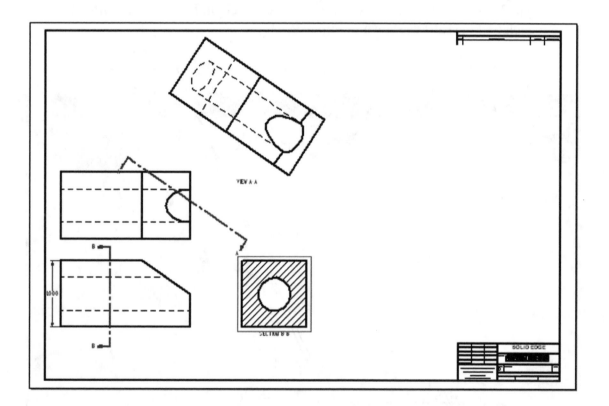

1. Start Solid Edge by referring to "Chapter 1 Getting Started".

2. After Solid Edge is running, open the .par file that was created in Chapter three. Move the cursor to the upper left corner of the screen and left click on the "Open" icon as shown in Figure 1.

Figure 1   **Left Click Here**

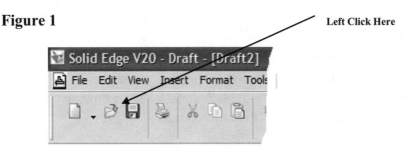

3. The Open File dialog box will appear. Locate the drawing that was created in Chapter three. Left click on the drop down arrow to the right of the text "Files of type:". A drop down menu will appear. Left click on **Draft documents.** Left click on **Part1.dft** as shown in Figure 2.

Figure 2   **Left Click Here**

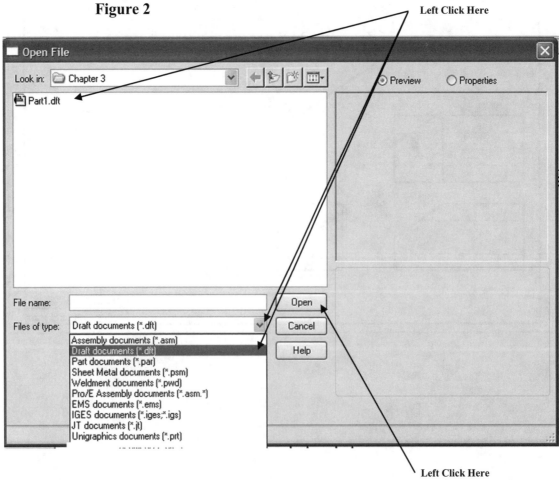

4. Left click on **Open**.

5.   After the .dft file is open, move the views closer to each other to provide additional room on the drawing. Start by moving the cursor over the top view. A red box will appear around the view. Left click (holding the left mouse button down) on the red box and drag the view down closer to the front view as shown in Figure 3.

Figure 3   Left Click Here, Hold Down and Drag

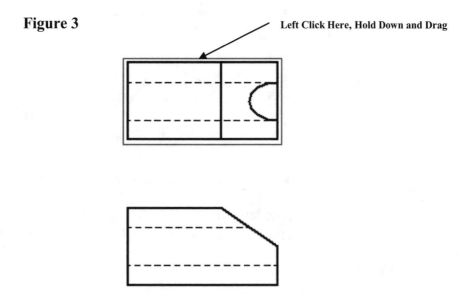

6.   Move the side view closer to the front view. Start by moving the cursor over the side view. A red box will appear around the view. Left click (holding the left mouse button down) on the red box and drag the view closer to the front view as shown in Figure 4.

Figure 4   Left Click Here, Hold Down and Drag

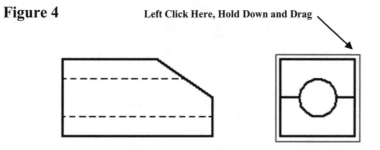

7.   Move the cursor over the isometric view causing the red box to appear and left click once. The red box will become purple dashed lines as shown in Figure 5.

Figure 5

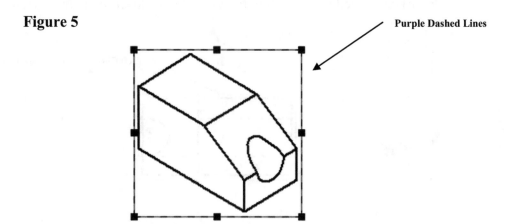

Purple Dashed Lines

8.   Move the cursor to the upper left portion of the screen and left click on **Edit**. A drop down menu will appear. Left click on **Delete** as shown in Figure 6.

Figure 6

Left Click Here

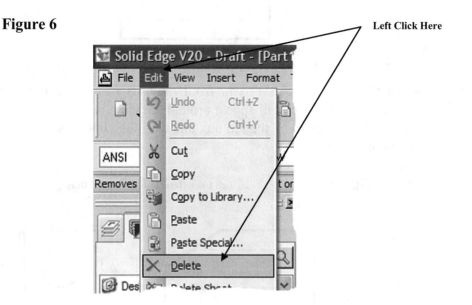

9.   A confirmation dialog box will appear. Left click on **Yes** as shown in Figure 7.

Figure 7

Left Click Here

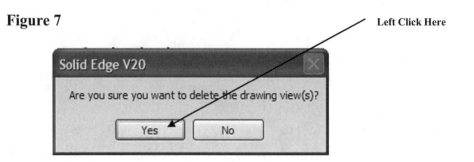

10. There will now be more room to work. Your screen should look similar to Figure 8.

Figure 8

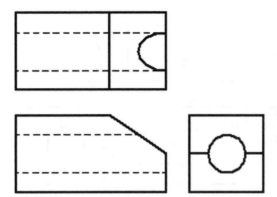

11. Move the cursor to the upper left portion of the screen and left click on **Auxiliary View** as shown in Figure 9.

Figure 9

Left Click Here

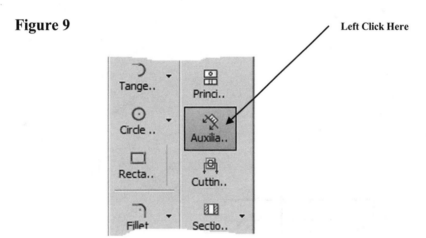

12. Move the cursor to the wedge line and left click as shown in Figure 10.

Figure 10

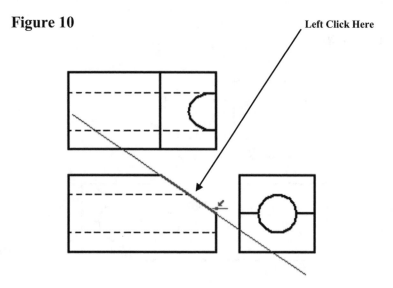

Left Click Here

13. Solid Edge will create an auxiliary view from the selected surface. The view will be attached to the cursor as shown in Figure 11.

Figure 11

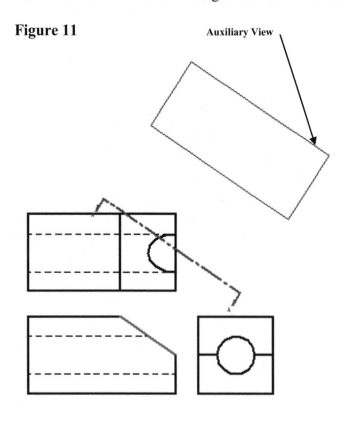

Auxiliary View

14. Drag the cursor away from the other views and left click once. Your screen should look similar to Figure 12.

Figure 12

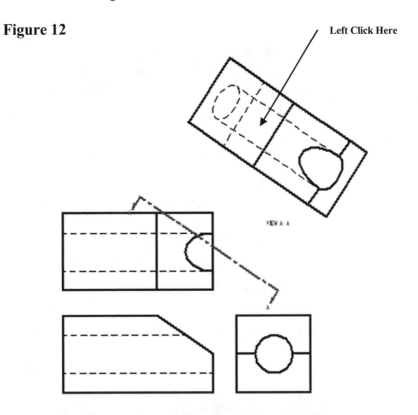

15. Move the cursor to the side view. A red box will appear as shown in Figure 13. Left click once causing the box to become purple with dashed lines.

Figure 13

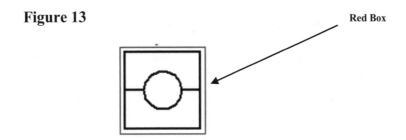

16.   Move the cursor to the upper left portion of the screen and left click on **Edit**. A drop down menu will appear. Left click on **Delete** as shown in Figure14.

Figure 14

Left Click Here

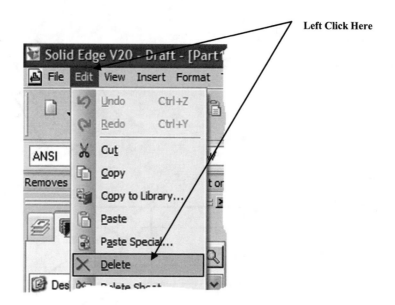

17.   A confirmation dialog box will appear. Left click on **Yes** as shown in Figure 15.

Figure 15

Left Click Here

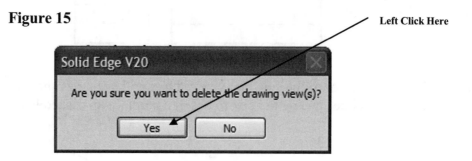

18.   The side view of the part will be deleted. Your screen should look similar to Figure 16.

Figure 16

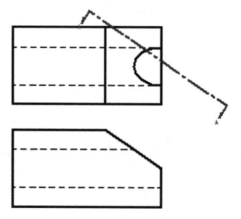

19. Move the cursor to the upper left portion of the screen and left click on **Cutting Plane** as shown in Figure 17.

Figure 17

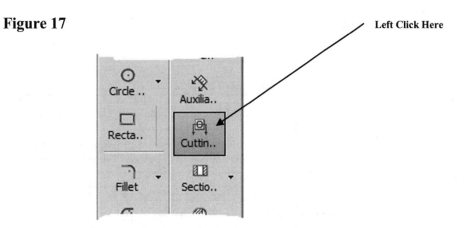

Left Click Here

20. Move the cursor over the front view causing a red box to appear around the part. Left click once as shown in Figure 18.

Figure 18

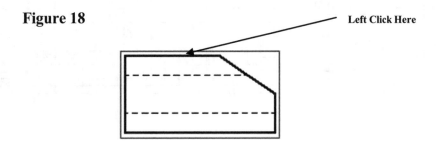

Left Click Here

21. Solid Edge will move into an area where a cutting plane line can be drawn. Your screen should look similar to Figure 19.

Figure 19

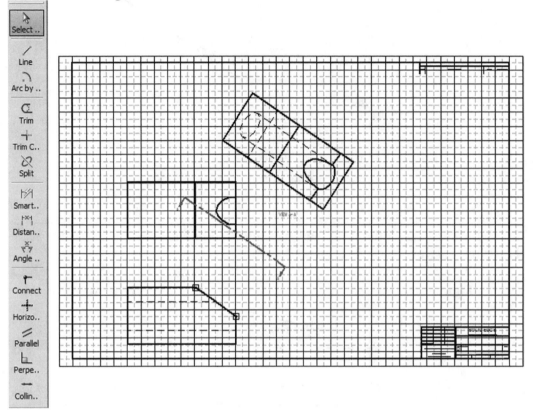

22. Move the cursor to the upper left portion of the screen and left click on **Line** as shown in Figure 20.

Figure 20

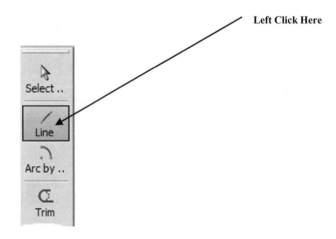

200

23. Move the cursor above the top edge of the part towards the center and left click once. While the line is still attached move the cursor down past the bottom edge of the part and left click once as shown in Figure 21.

Figure 21

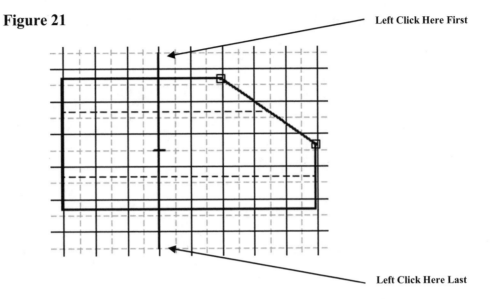

Left Click Here First

Left Click Here Last

24. Move the cursor to the upper left portion of the screen and left click on **Finish** as shown in Figure 22.

Figure 22

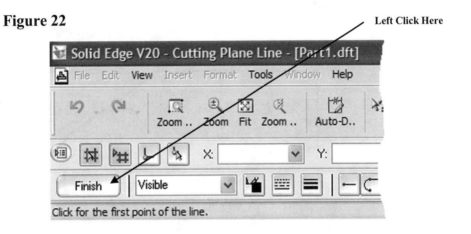

Left Click Here

25. The direction of the cutting plane line will be attached to the cursor. Move the cursor to the left side of the front view causing the arrows to face towards the left and left click once as shown in Figure 23.

Figure 23

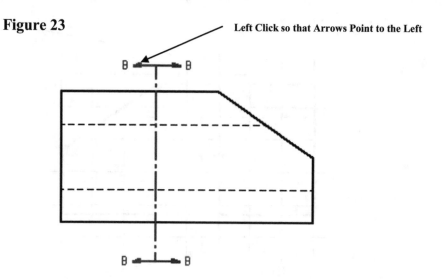

Left Click so that Arrows Point to the Left

26. Your screen should look similar to Figure 24.

Figure 24

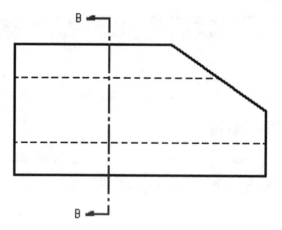

27.   Move the cursor to the upper left portion of the screen and left click on **Section View** as shown in Figure 25.

Figure 25

Left Click Here

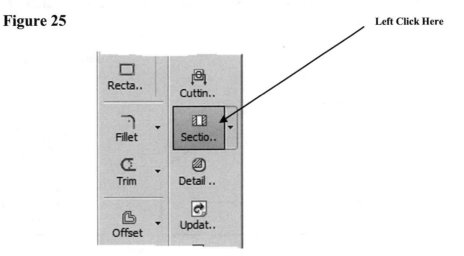

28.   Move the cursor over the cutting plane line causing it to turn red and left click once. The cutting plane line will turn purple and the side view of the part will appear attached to the cursor as shown in Figure 26.

Figure 26

Left Click Here

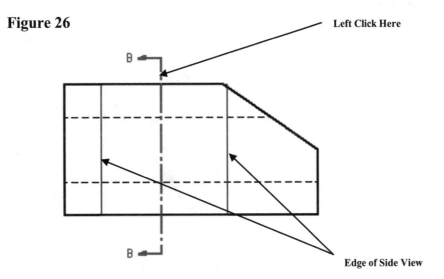

Edge of Side View

203

29. Drag the cursor to the right. The section view will be attached to the cursor. Place the section view to the side of the front view and left click as shown in Figure 27.

Figure 27

Left Click Here

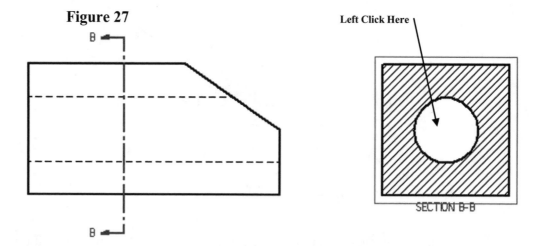

SECTION B-B

30. Move the cursor to the upper left portion of the screen and left click on **SmartDimension** as shown in Figure 28.

Figure 28

Left Click Here

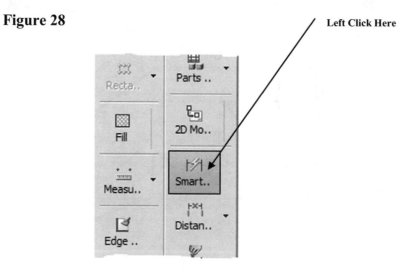

31.   Move the cursor over the left side vertical line until it turns red as shown in
       Figure 29. Select the line by left clicking anywhere on the line **or** on each of the
       end points. The dimension will be attached to the cursor.

Figure 29

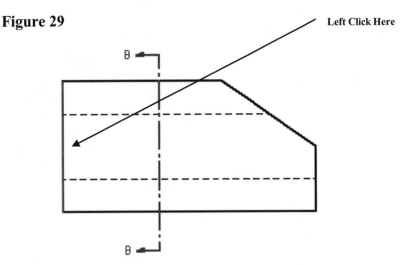

32.   Move the cursor to where the dimension will be placed and left click once as
       shown in Figure 30.

Figure 30

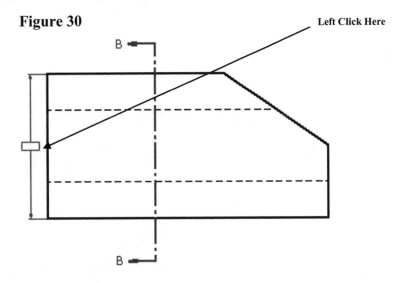

33.   Your screen should look similar to Figure 31.

Figure 31

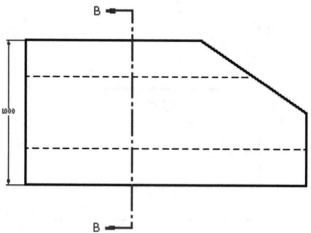

34.   Finish dimensioning the part to your own satisfaction. When the part is satisfactorily dimensioned, save the file to a location where it can be easily retrieved.

35.   To edit the text height, move the cursor over the dimension, (after pressing the **ESC** key once or twice) and right click as shown in Figure 32.

Figure 32

Right Click Once

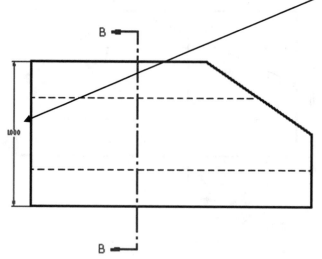

36. A pop up menu will appear. Left click on **Properties** as shown in Figure 33.

Figure 33

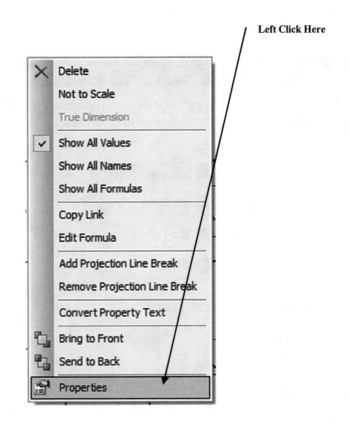

37.   The Dimension Properties dialog box will appear as shown in Figure 34.

Figure 34

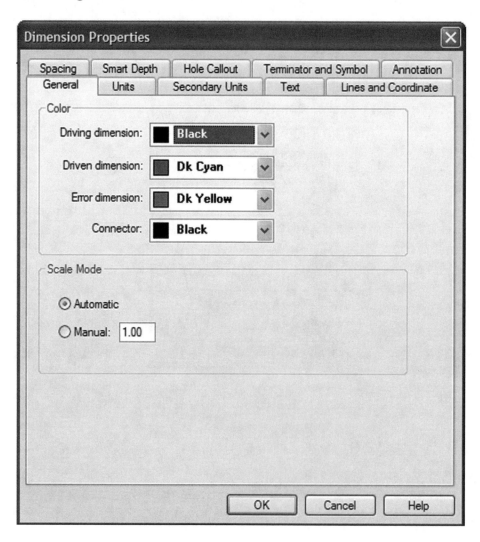

38. Left click on the **Text** tab. Enter **.250** for Font Size as shown in Figure 35.

Figure 35

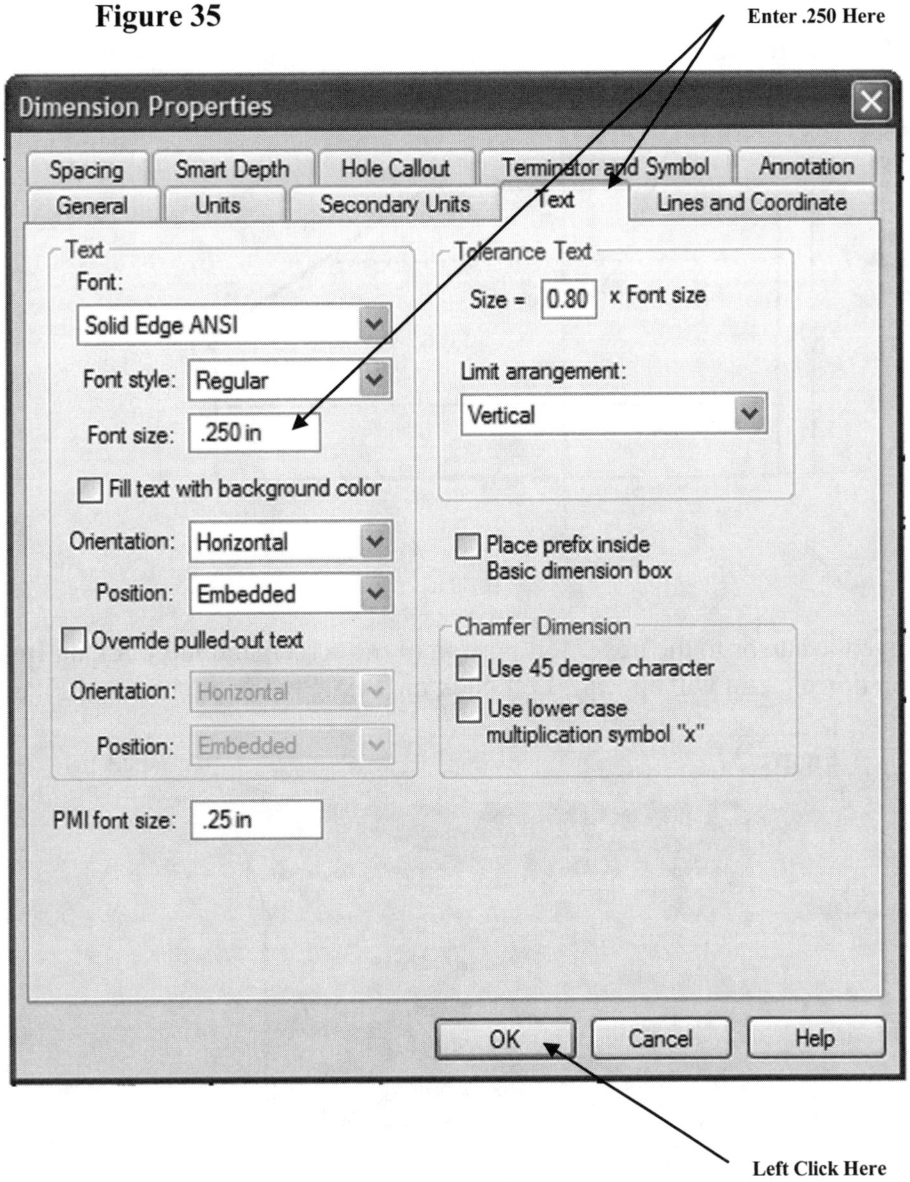

Enter .250 Here

Left Click Here

39. Left click on **OK**.

40. Your screen should look similar to Figure 36.

Figure 36

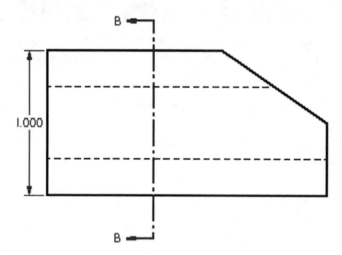

41. Move the cursor to the upper left portion of the screen and left click on **Insert**. A drop down menu will appear. Left click on **Text Profile** as shown in Figure 37.

Figure 37   Left Click Here

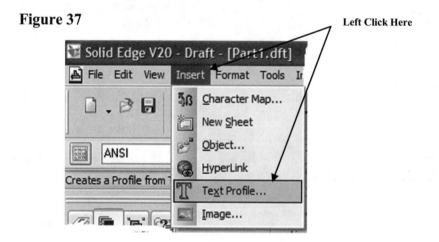

42.   The Text dialog box will appear. Type your name as shown in Figure 38.

Figure 38

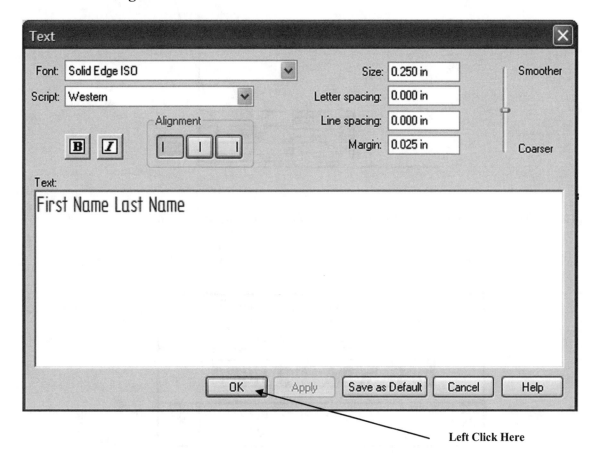

Left Click Here

43.   Left click on **OK**.

44.   A small box will be attached to the cursor as shown in Figure 39.

Figure 39   Attached to Cursor

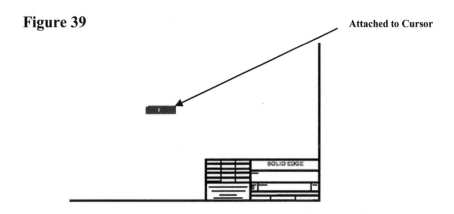

45. Move the cursor to the title block location and left click as shown in Figure 40.

Figure 40

Left Click Here

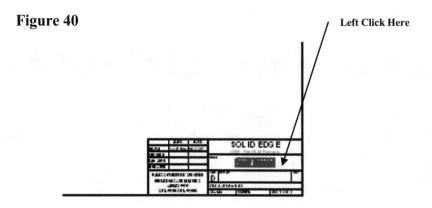

46. Text will appear in the title block as shown in Figure 41.

Figure 41

Text that was Typed

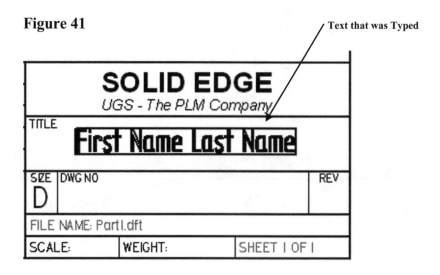

47.   If the text needs to be moved, move the cursor over the text causing a red box to
       appear as shown in Figure 42.

Figure 42

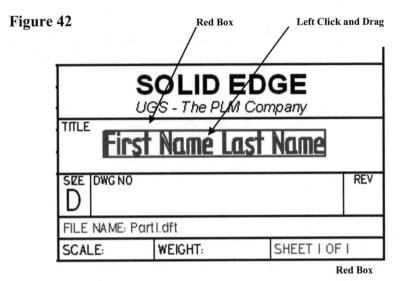

48.   While the text is highlighted, left click (holding the left mouse button down) and
       drag the text to the desired location. After the text is in the desired location,
       release the left mouse button.

49.   Your screen should look similar to Figure 43.

Figure 43

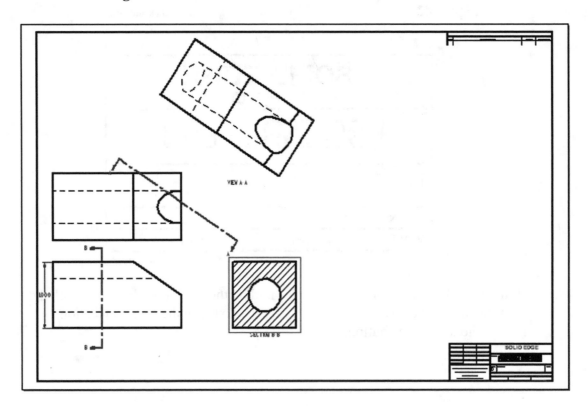

50.   Save the current sheet where it can be easily retrieved.

Drawing Activities

Create Section View Drawings for the following:

Problem 1

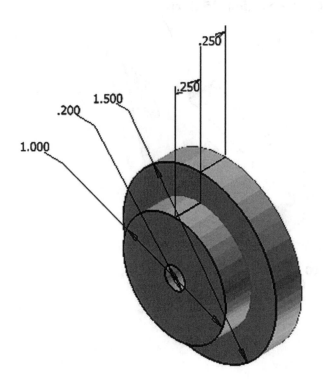

Problem 2

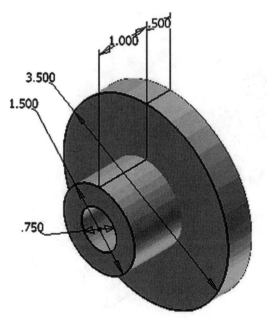

Create Auxiliary View Drawings for the following:

Problem 3

Extrude Center Section .25 Deep

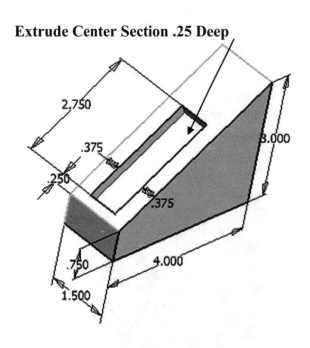

Problem 4

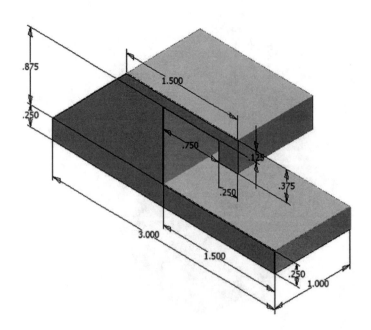

Create Section View Drawings for the following:

Problem 5

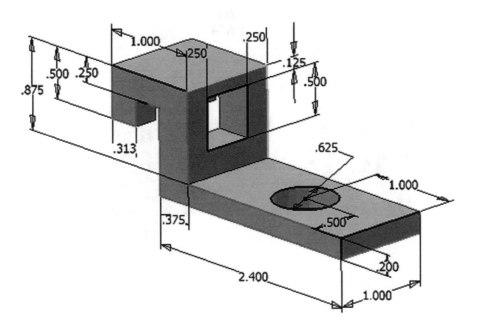

Problem 6

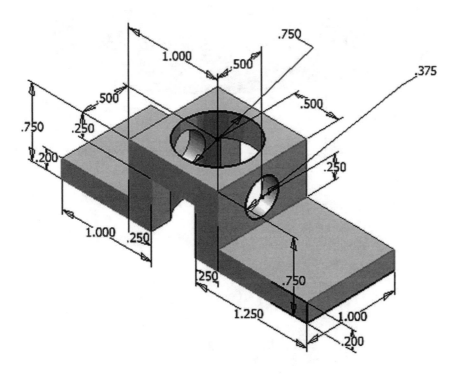

Create Section View Drawings for the following:

Problem 7

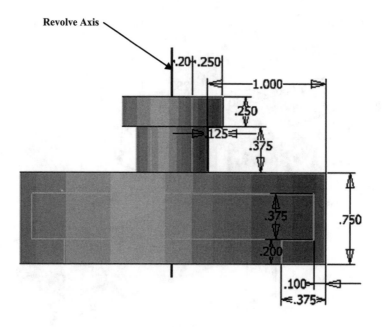

Problem 8

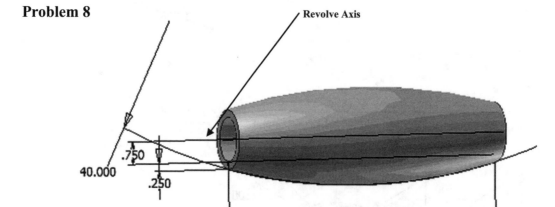

Chapter 5 Learning To Edit Existing Solid Models

Objectives:

- Design a simple part
- Learn to use the Pattern Along a Curve command
- Learn to edit a part using the Pattern Along a Curve command
- Edit the part using the Sketch command
- Edit the part using the Protrusion command
- Edit the part using the Chamfer command

Chapter 5 includes instruction on how to design and edit the part shown below.

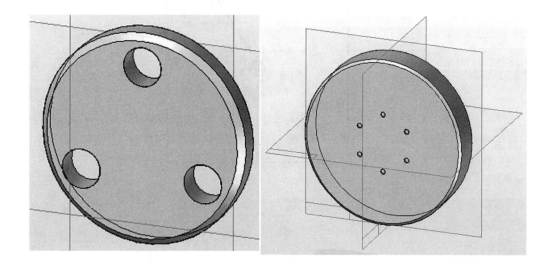

1. Start Solid Edge by referring to "Chapter 1 Getting Started".

2. After Solid Edge is running, begin a new sketch.

3. Move the cursor to the upper left corner of the screen and left click on **Circle** as shown in Figure 1.

Figure 1

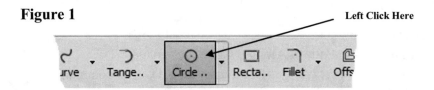

4. Move the cursor to the center of the screen and left click once. This will be the center of the circle as shown in Figure 2.

Figure 2

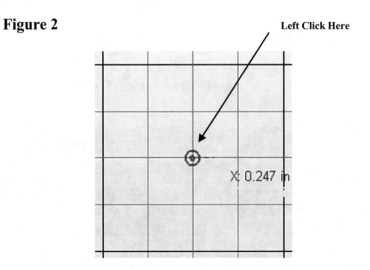

5.   Move the cursor to the right and left click once as shown in Figure 3.

Figure 3

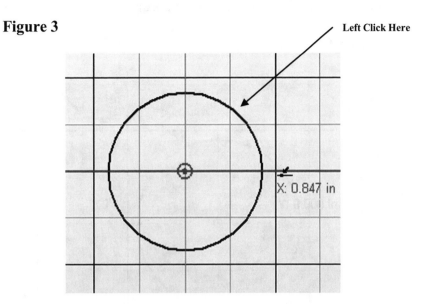

6.   Move the cursor to the middle left portion of the screen and left click on **SmartDimension** as shown in Figure 4.

Figure 4

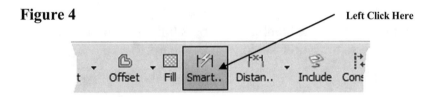

7.   Move the cursor over the edge (not center) of the circle until it turns red. Left click once as shown in Figure 5. The dimension box will be attached to the cursor.

Figure 5

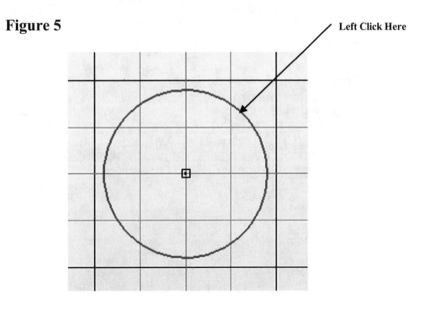

221

8.   Move the cursor to where the dimension will be placed and left click once as shown in Figure 6. The dimension will appear in the box.

Figure 6

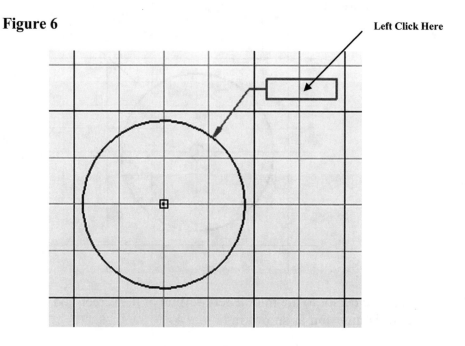

9.   While the dimension is highlighted type **2.000** in the dimension box as shown in Figure 7. Press the **Enter** key on the keyboard.

Figure 7

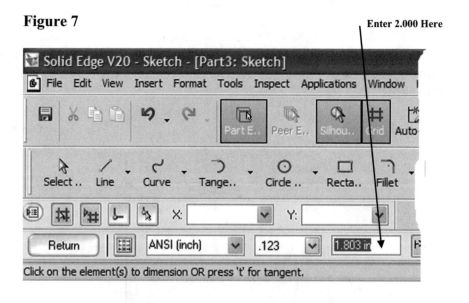

10. The dimension of the circle will become 2.00 inches as shown in Figure 8.

Figure 8

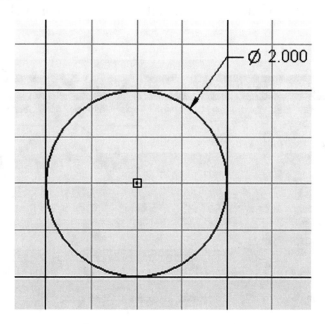

11. To view the entire drawing, it may be necessary to move the cursor to the upper right portion of the screen and left click once on the "Fit" icon as shown in Figure 9.

Figure 9

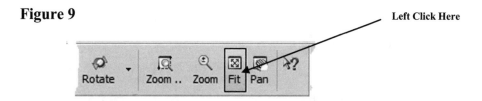

12. The drawing will "fill up" the entire screen. If the drawing is still too large, left click on the "Zoom" icon as shown in Figure 10. After selecting the "Zoom" icon, hold the left mouse button down and drag the cursor up and down to achieve the desired view of the sketch.

Figure 10

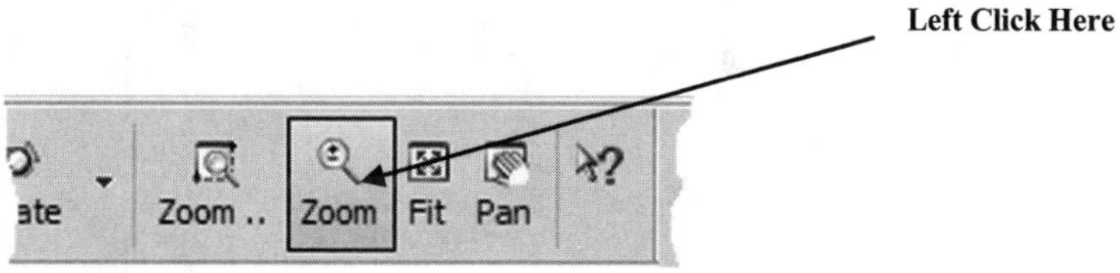

13. After the sketch is complete it is time to extrude the sketch into a solid.

14. After you have verified that no commands are active, move the cursor to the upper left portion of the screen and left click on **Return** as shown in Figure 11.

Figure 11   Left Click Here

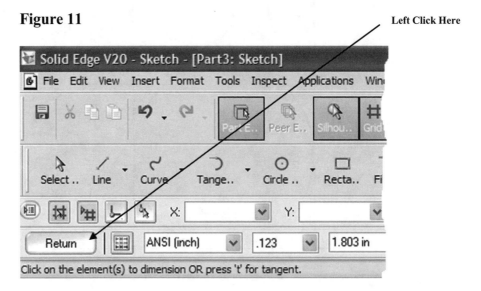

15. Solid Edge is now out of the Sketch area and into the Model area. Notice that the commands at the top of the screen are now different. To work in the Model area a sketch must be present and have no opens (non-connected lines). If there are any opens in the sketch an error message may appear. Your screen should look similar to Figure 12.

Figure 12

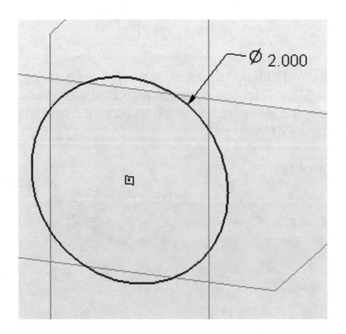

16.   Move the cursor to the upper left portion of the screen and left click on **Protrusion** as shown in Figure 13.

Figure 13   Left Click Here

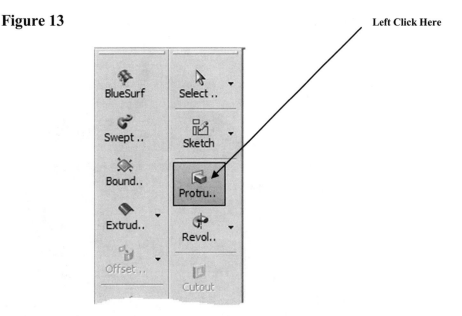

17.   A preview of the extrusion will be displayed. If Solid Edge gave you an error message, there are opens (non-connected lines) somewhere on the sketch. Check each intersection for opens by using the **Extend** and **Trim** commands.

18.   Move the cursor over any portion of the sketch causing the entire profile to turn red. Left click as shown in Figure 14. The sketch will become yellow. After the sketch becomes yellow, right click once.

Figure 14   Left Click/Right Click Here

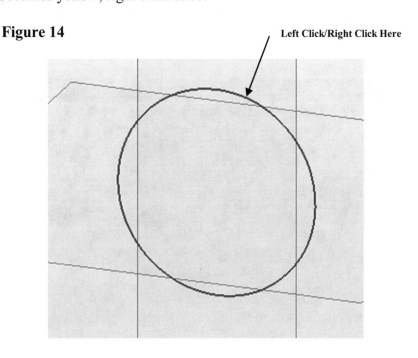

19. The sketch will become three dimensional as shown in Figure 15. The extruded surface will be attached to the cursor.

Figure 15

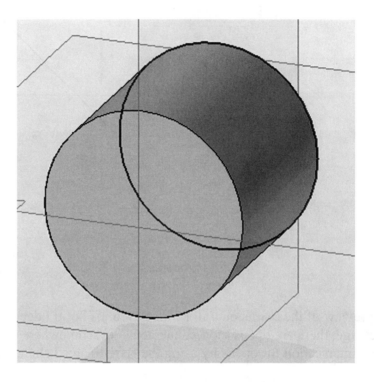

20. Move the cursor to the upper middle portion of the screen and type **.250** next to the text "Distance" and **.250** next to the text "Step" as shown in Figure 16. Left click anywhere around the sketch.

Figure 16

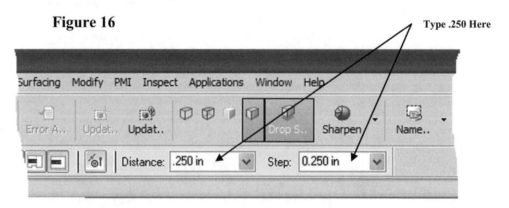

21.   Move the cursor to the upper left portion of the screen and left click on **Finish** as shown in Figure 17.

Figure 17

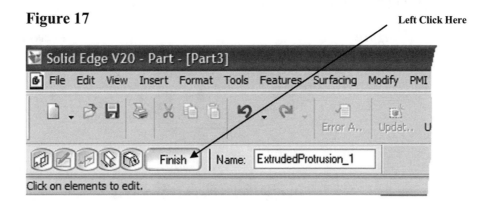

22.   Your screen should look similar to Figure 18.

Figure 18

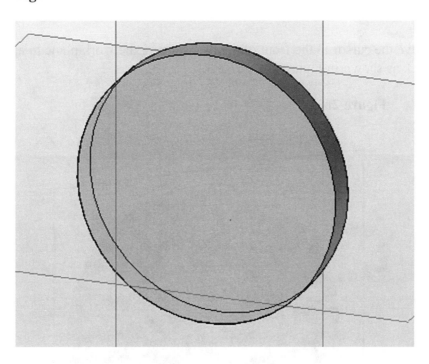

23. Move the cursor to the upper left portion of the screen and left click on **Sketch** as shown in Figure 19.

Figure 19

Left Click Here

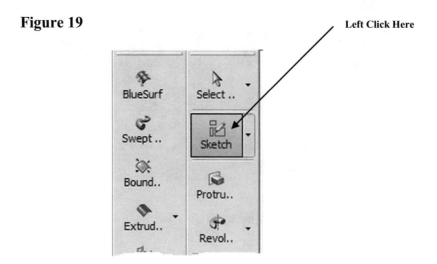

24. Move the cursor to the front of the part causing the work plane to appear and left click as shown in Figure 20.

Figure 20

Left Click Here

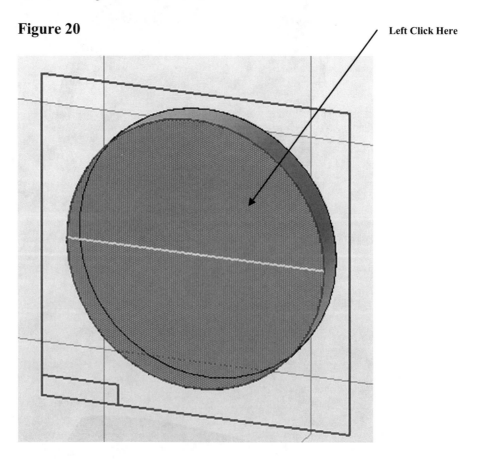

25. Solid Edge will start a new sketch on the selected surface as shown in Figure 21.

Figure 21

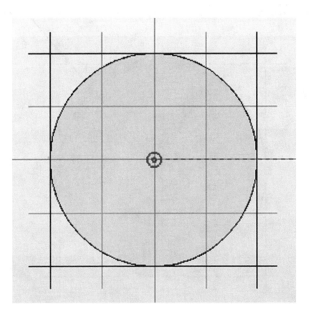

26. Move the cursor to the upper left corner of the screen and left click on **Circle** as shown in Figure 22.

Figure 22

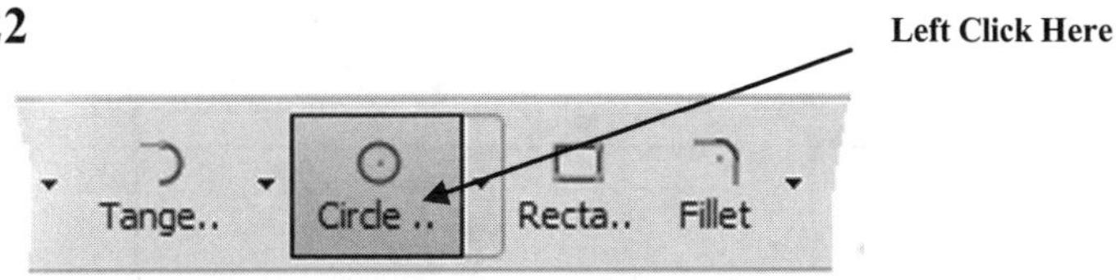

Left Click Here

Tange..   Circle ..   Recta..   Fillet

27.   Move the cursor to the center of the circle. A red dot will appear as shown in Figure 23.

Figure 23

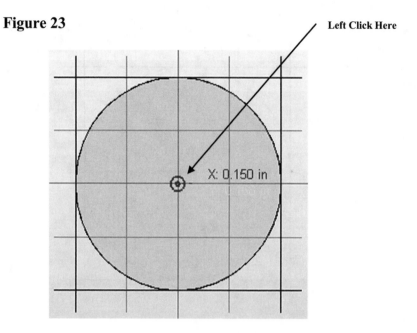

28.   After the red dot appears, move the cursor straight up along the line coming out of the center of the part and left click once as shown in Figure 24.

Figure 24

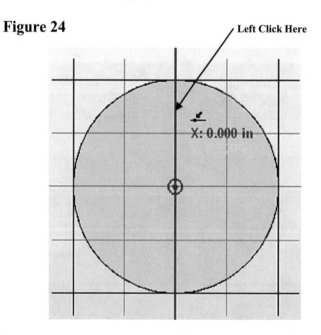

29. Move the cursor to the side. A circle will form. Left click once as shown in Figure 25.

Figure 25

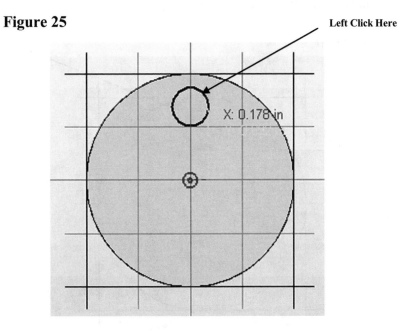

Left Click Here

X: 0.178 in

30. Move the cursor to the middle left portion of the screen and left click on **SmartDimension** as shown in Figure 26.

Figure 26

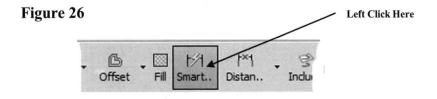

Left Click Here

Offset   Fill   Smart..   Distan..   Indu

31. Move the cursor over the edge (not center) of the circle until it turns red as shown in Figure 27. Select the circle by left clicking anywhere on the edge. The dimension will be attached to the cursor.

Figure 27

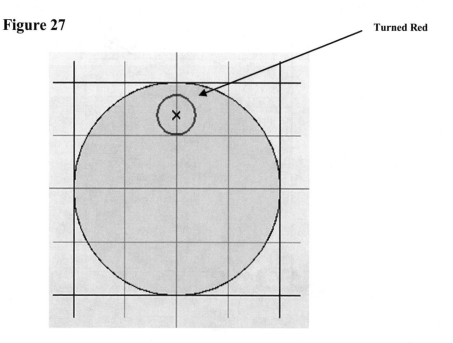

32. Move the cursor to where the dimension will be placed and left click once as shown in Figure 28.

Figure 28

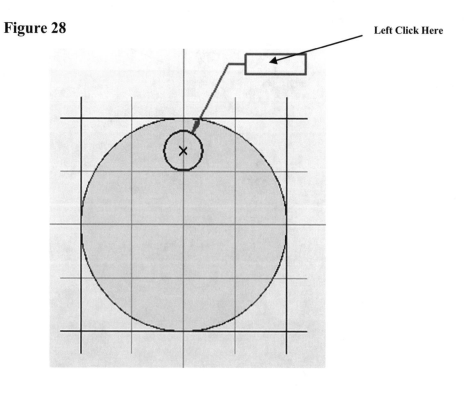

33. Move the cursor to the upper middle portion of the screen and type **.375** in the dimension box as shown in Figure 29. Press the **Enter** key on the keyboard.

Figure 29

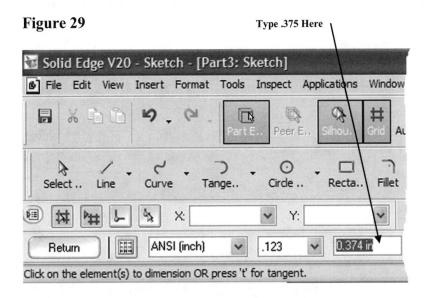

34. The dimension of the circle will become .375 inches as shown in Figure 30. Use the Zoom icons to zoom out if necessary.

Figure 30

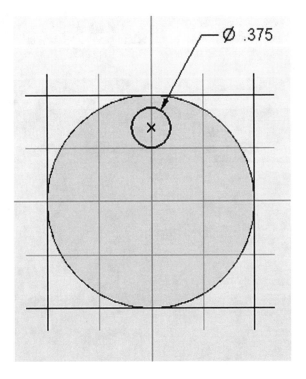

35. After you have verified that no commands are active, move the cursor to the upper left portion of the screen and left click on **Return** as shown in Figure 31.

Figure 31

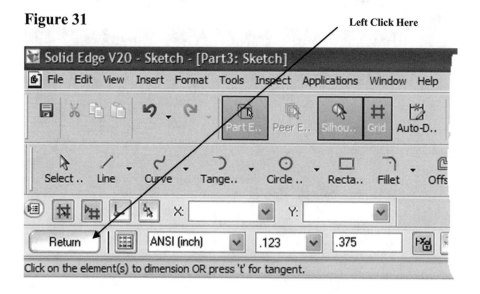

36. Move the cursor to the upper left portion of the screen and left click on **Finish** as shown in Figure 32.

Figure 32

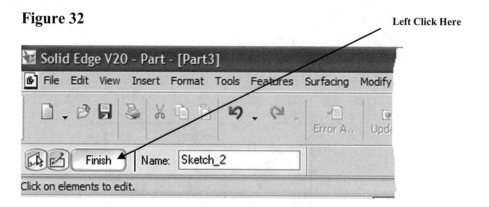

37. Solid Edge is now out of the Sketch area and into the Model area. Notice that the commands at the top of the screen are now different. To work in the Model area a sketch must be present and have no opens (non-connected lines). If there are any opens in the sketch an error message will appear. Your screen should look similar to Figure 33.

Figure 33

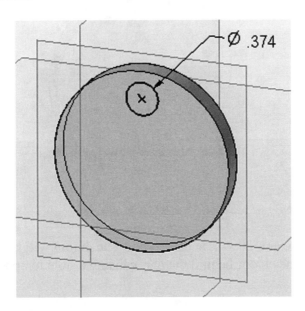

38. Move the cursor to the upper left portion of the screen and left click on **Cutout** as shown in Figure 34.

Figure 34

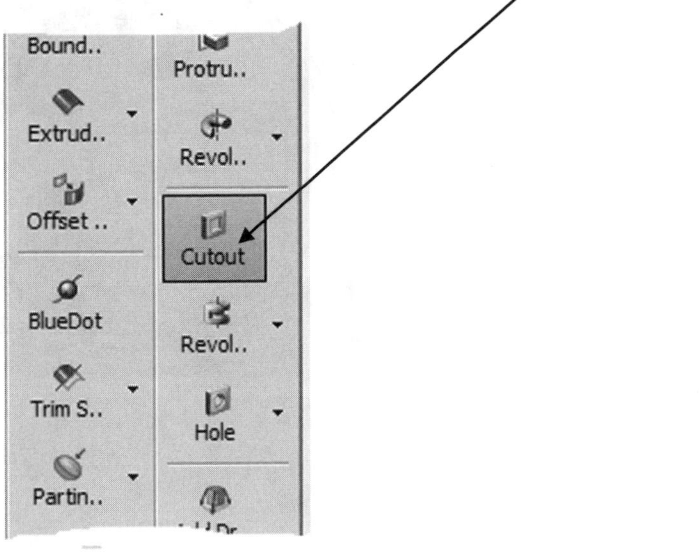

39. Move the cursor over the circle causing it to turn red. Left click once then right click once as shown in Figure 35.

Figure 35

Left Click Then Right Click Here

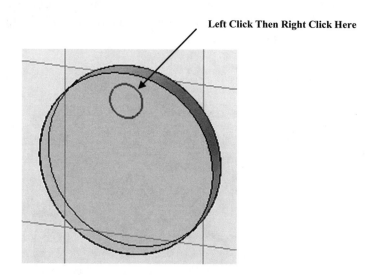

40. A preview of the cutout will be displayed. The cutout will be attached to the cursor. Move the cursor behind the part causing a hole to appear as shown in Figure 36.

Figure 36

Cutout is Attached to Cursor

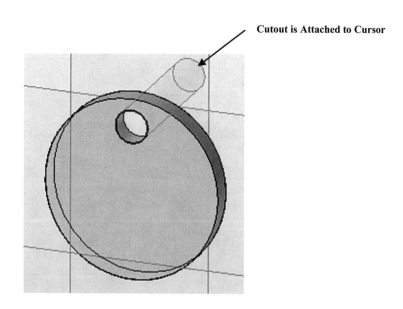

41.   Move the cursor to the upper middle portion of the screen and Enter **.250** next to the text "Distance" as shown in Figure 37. Press the **Enter** key on the keyboard.

Figure 37

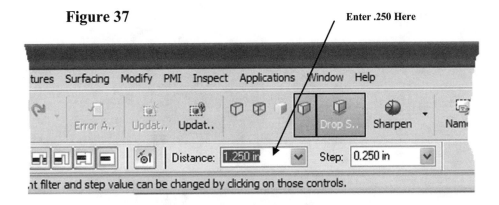

42.   Move the cursor behind the part and left click once as shown in Figure 38.

Figure 38   Left Click Here

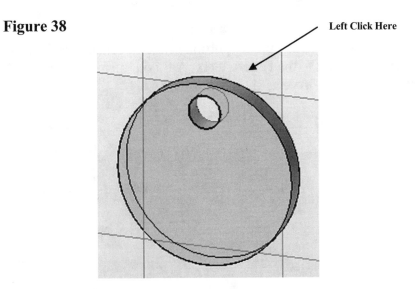

43.   Move the cursor to the upper left portion of the screen and left click on **Finish** as shown in Figure 39.

Figure 39   Left Click Here

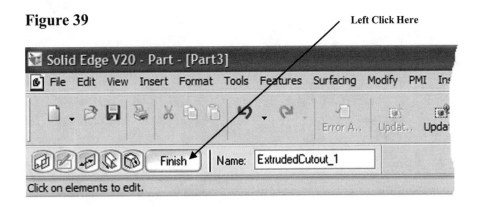

44. Your screen should look similar to Figure 40. You may have to use the zoom out command to view the entire part.

Figure 40

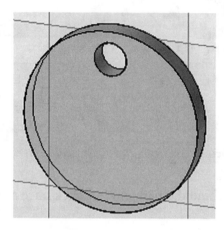

45. Move the cursor to the middle portion of the screen and left click on the drop down arrow next to the Pattern icon. A fly out menu will appear. Left click on **Pattern Along Curve** as shown in Figure 41.

Figure 41

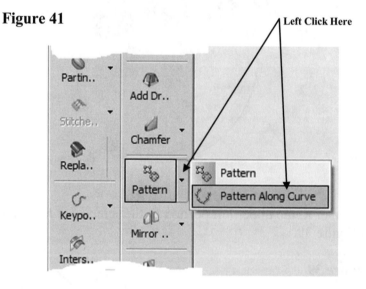

46. Move the cursor to the center of the hole causing the edges to turn red. Left click then right click as shown in Figure 42.

Figure 42

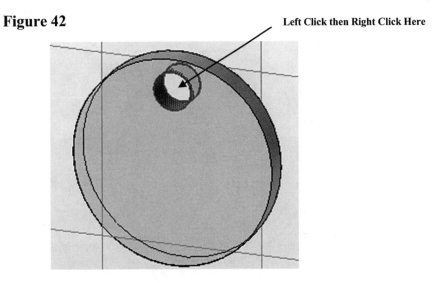

Left Click then Right Click Here

47. Move the cursor to the upper middle portion of the screen and type **3** next to the text "Count" as shown in Figure 43.

Figure 43

Type 3 Here

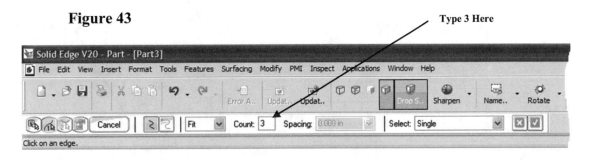

48. Move the cursor to the edge of the part causing it to turn red and left click once as shown in Figure 44.

Figure 44

Left Click Here

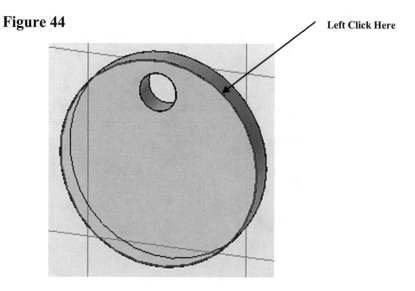

49. Move the cursor to the upper right portion of the screen and left click on the green checkmark as shown in Figure 45.

Figure 45

Left Click Here

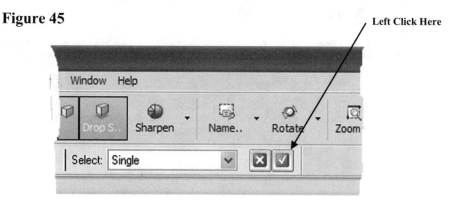

50. An arrow will appear attached to the cursor as shown in Figure 46.

Figure 46   **Attached to Cursor**

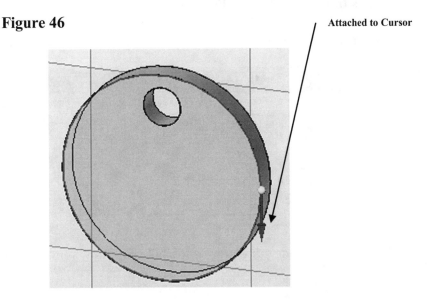

51. Move the cursor down causing the arrow to point downward and left click once. Solid Edge will provide a preview of the hole pattern as shown in Figure 47.

Figure 47   **Left Click Here**

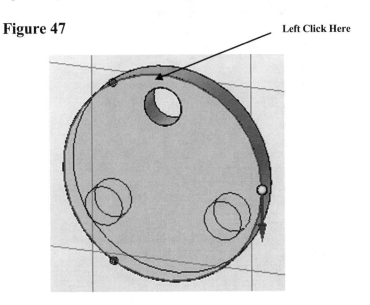

52.   Move the cursor to the left portion of the screen and left click **Full** and **Curve Position**. Left click on **Preview** as shown in Figure 48.

Figure 48

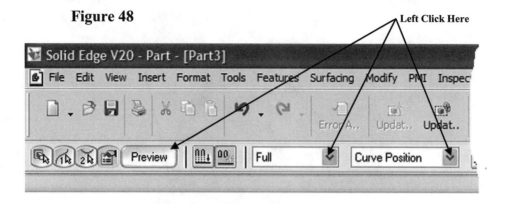

53.   Solid Edge will provide a preview of the holes as shown in Figure 49.

Figure 49

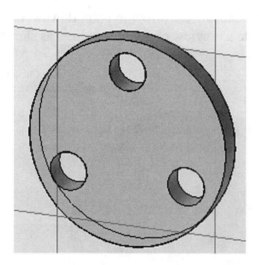

54.   Move the cursor to the upper left portion of the screen and left click on **Finish** as shown in Figure 50.

Figure 50

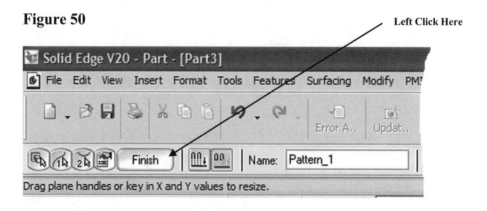

55. Your screen should look similar to Figure 51.

Figure 51

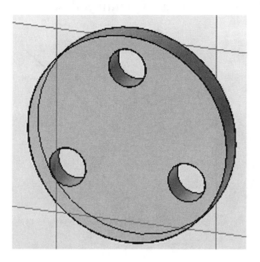

56. Move the cursor to the middle left portion of the screen and left click on the drop down arrow next to the Round icon. A fly out menu will appear. Left click on **Chamfer** as shown in Figure 52.

Figure 52

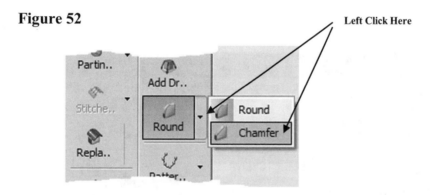

57. Enter **.0625** for the Setback as shown in Figure 53.

Figure 53

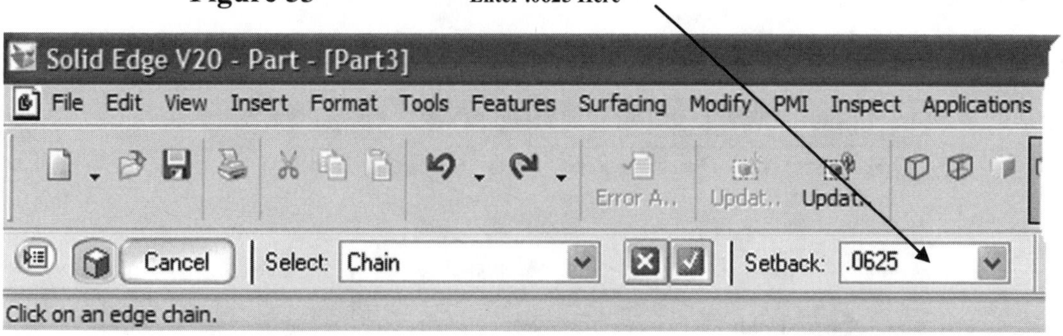

243

58. Move the cursor over the edge of the part causing it to turn red as shown in Figure 54. Left click once. The edge will turn yellow.

Figure 54

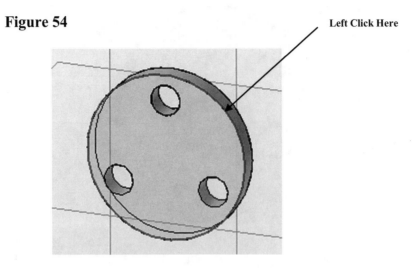

Left Click Here

59. Move the cursor to the upper middle portion of the screen and left click on the green checkmark as shown in Figure 55.

Figure 55

Left Click Here

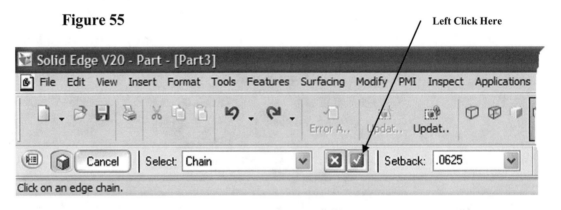

60. Move the cursor to the upper left portion of the screen and left click on **Finish** as shown in Figure 56.

Figure 56

Left Click Here

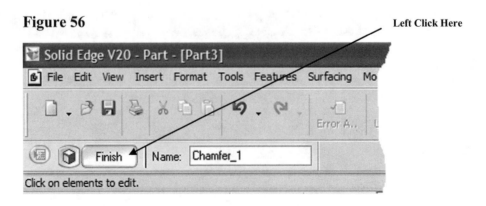

61. Your screen should look similar to Figure 57.

Figure 57

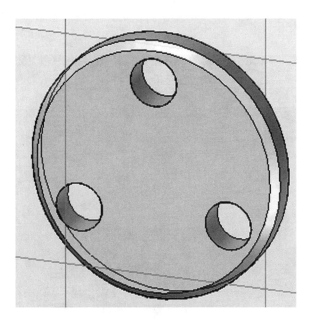

62. If for some reason a change needs to be made to this part, it can be accomplished by editing either a sketch or a definition located in the part history tree at the upper left corner of the screen as shown in Figure 58.

Figure 58

Part History Tree Location

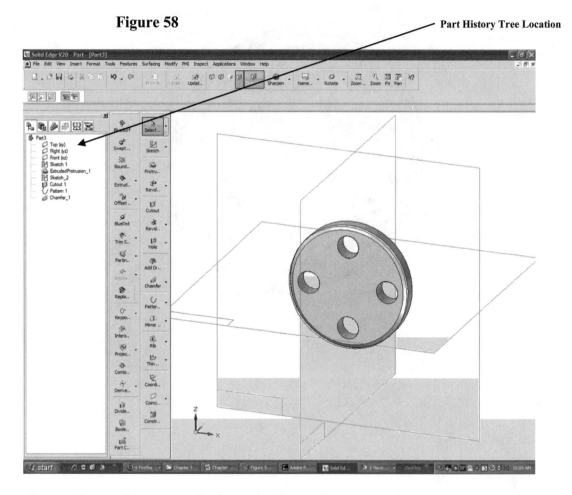

63. The part history tree is shown in Figure 59.

Figure 59

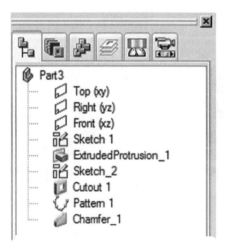

64. Move the cursor over the text "Sketch 1". A red box will appear around the text as shown in Figure 60.

Figure 60

Highlighted Text

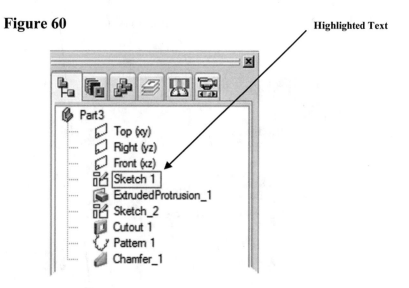

65. The original sketch will also appear as shown in Figure 61.

Figure 61

Original Sketch

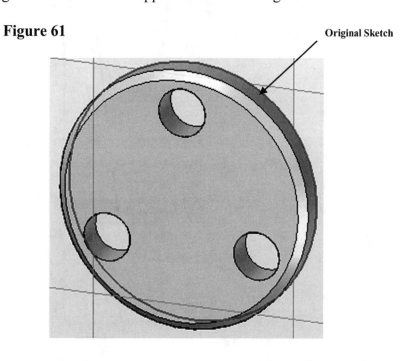

66.   Right click on **Sketch 1**. A pop up menu will appear. Left click on **Edit Profile** as shown in Figure 62.

Figure 62

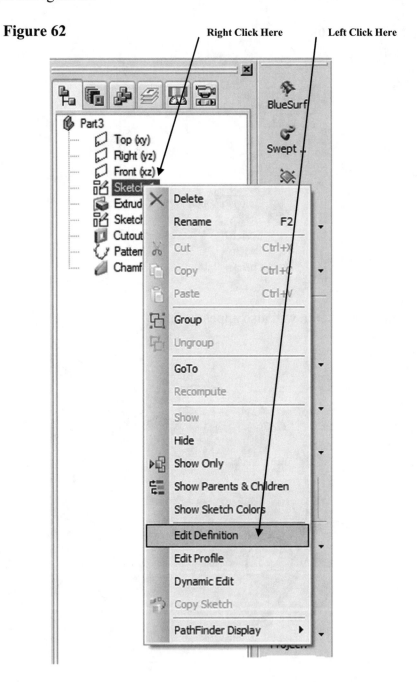

67. The original sketch will appear as shown. Single left click on the overall dimension as shown in Figure 63.

Figure 63

Single Left Click Here

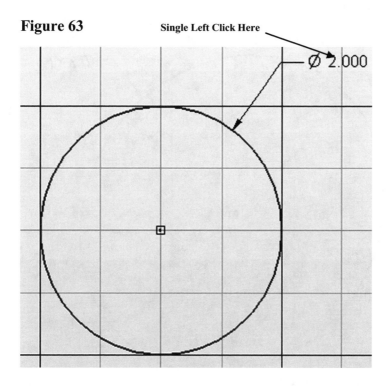

68. Modify the diameter of the part by entering **4.00** in the dimension box as shown in Figure 64.

Figure 64

Enter 4.00 Here

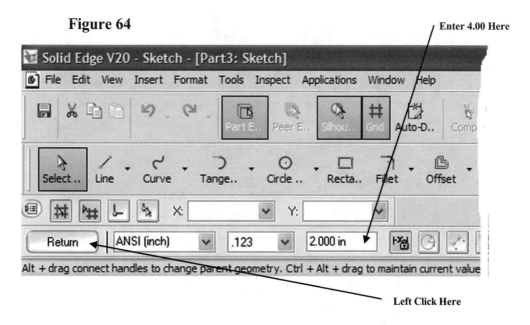

Left Click Here

69. Left click on **Return**.

70.   The diameter of the part will increase to 4.00 in as shown in Figure 65.

Figure 65

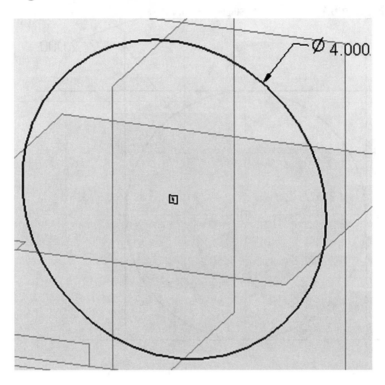

71.   Move the cursor to the upper left portion of the screen and left click on **Finish** as shown in Figure 66.

Figure 66

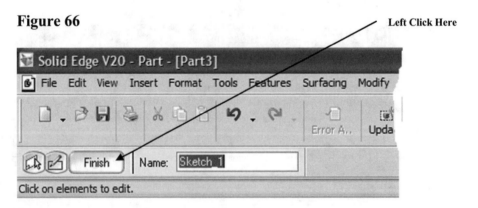

72.   Solid Edge will automatically update the part to reflect the changes made in the
sketch. It will not be necessary to extrude the part again. Your screen should
look similar to Figure 67.

Figure 67

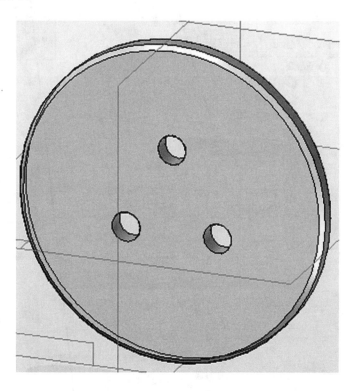

73.   Move the cursor to the part tree. Right click once on **ExtrudedProtrusion_1**. A
       pop up menu will appear. Left click on **Edit Definition** as shown in Figure 68.

Figure 68

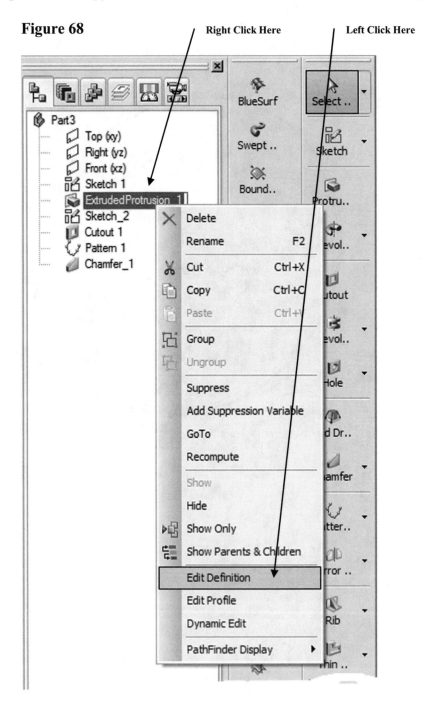

252

74. Solid Edge will return to the sketch area. Move the cursor over the dimension and single left click as shown in Figure 69.

Figure 69   Single Left Click Here

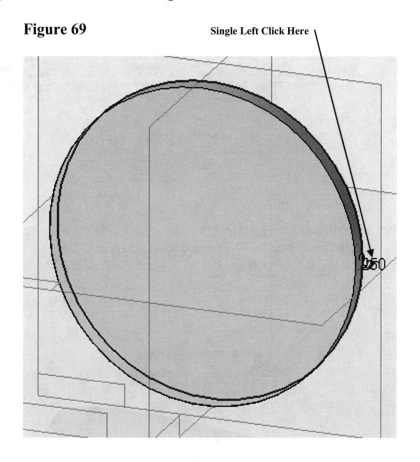

75. Move the cursor to the upper middle portion of the screen and enter **.750** in the dimension box as shown in Figure 70.

Figure 70   Enter .750 Here

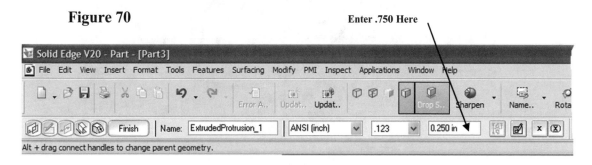

76. Move the cursor to the upper left portion of the screen and left click on **Finish** as shown in Figure 71.

Figure 71

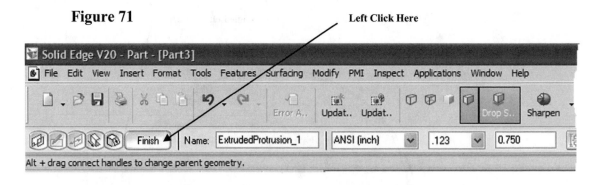

77. Solid Edge will automatically update the part without the need to repeat any of the steps that created the original part. Notice that the holes are no longer "thru" holes. Your screen should look similar to Figure 72.

Figure 72

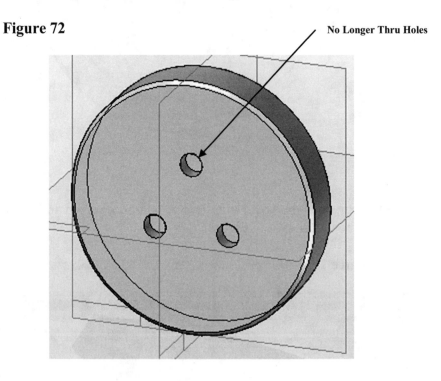

78. Move the cursor to the upper left portion of the screen where the part tree is located. Right click once on **Sketch 2**. A pop up menu will appear. Left click on **Edit Profile** as shown in Figure 73.

Figure 73

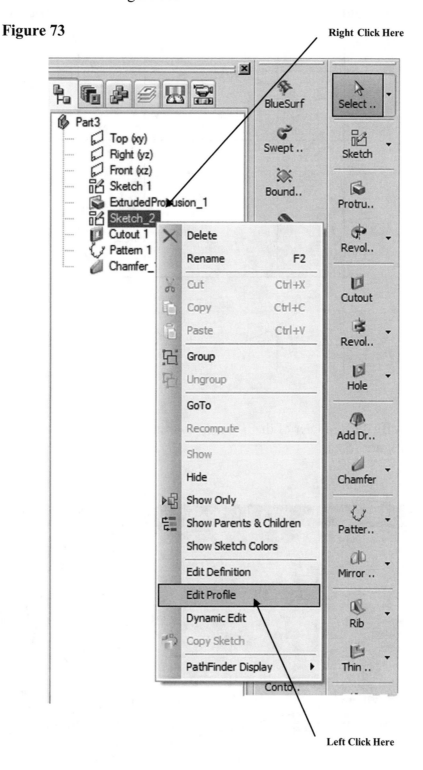

255

79. The original sketch will appear. Single left click on the overall dimension as shown in Figure 74.

Figure 74

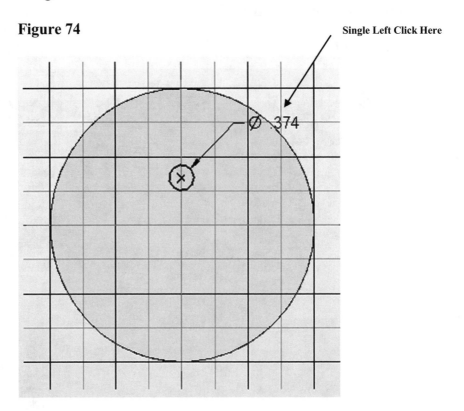

80. Modify the diameter of the holes by entering **.125** in the dimension box as shown in Figure 75.

Figure 75

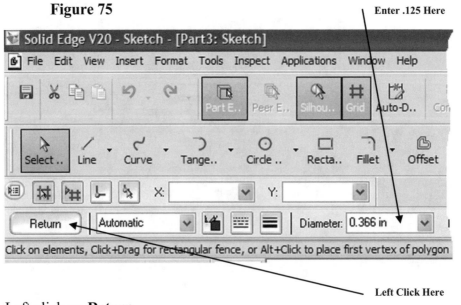

81. Left click on **Return**.

82. The diameter of the holes will be reduced to .125 as shown in Figure 76.

Figure 76

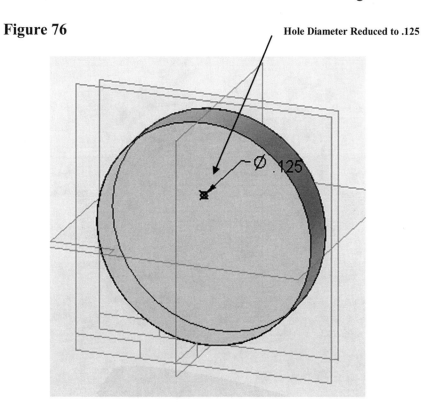

Hole Diameter Reduced to .125

83. Move the cursor to the upper left portion of the screen and left click on **Finish** as shown in Figure 77.

Figure 77

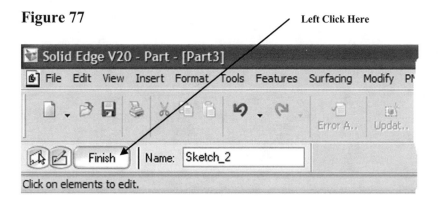

Left Click Here

84. Solid Edge will automatically update the part as shown in Figure 78.

Figure 78

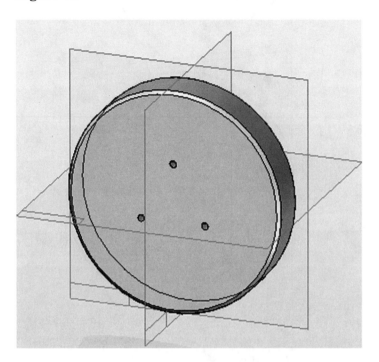

85. Move the cursor to the upper left portion of the screen where the part tree is located. Right click on **Cutout 1**. A pop up menu will appear. Left click on **Edit Definition** as shown in Figure 79.

Figure 79

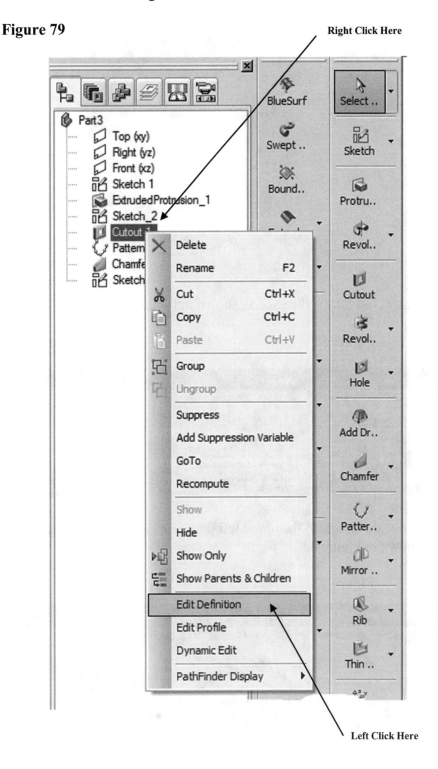

259

86. Move the cursor over the dimension and single left click as shown in Figure 80.

Figure 80

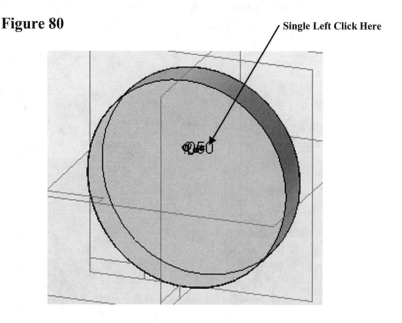

87. Enter **.750** for the distance as shown in Figure 81.

Figure 81

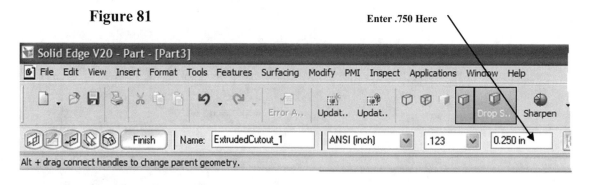

88. Move the cursor to the upper left portion of the screen and left click on **Finish** as shown in Figure 82.

Figure 82

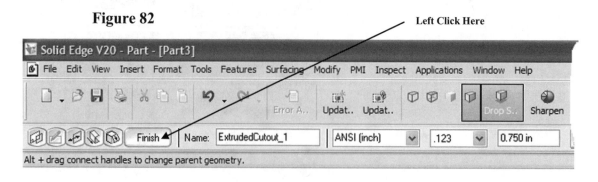

89. Solid Edge will automatically update the part. Notice that the holes are now "thru" holes as shown in Figure 83.

Figure 83

Thru Holes

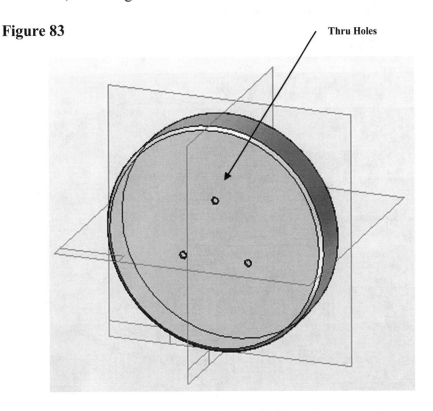

90. Move the cursor to the upper middle portion of the screen and left click on the drop down arrow next to the "Named Views" icon. A drop down menu will appear. Left click on **front** as shown in Figure 84.

Figure 84

Left Click Here

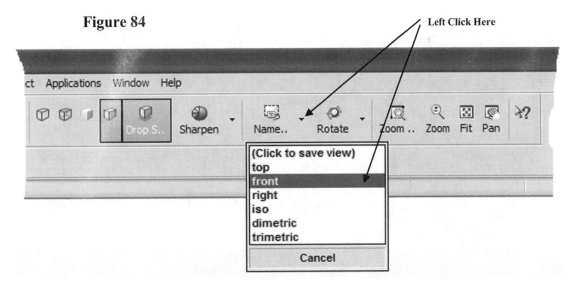

91.   A perpendicular view of the part will be displayed. Verify the holes are actually thru holes as shown in Figure 85.

Figure 85

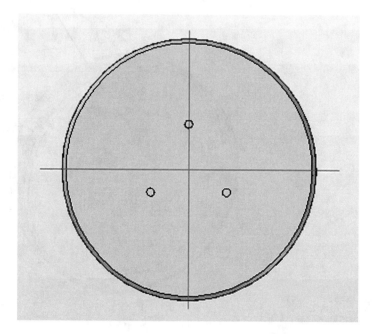

92.   Move the cursor to the upper left portion of the screen where the part tree is located. Right click on **Pattern_1**. A pop up menu will appear. Left click on **Edit Definition** as shown in Figure 86.

Figure 86

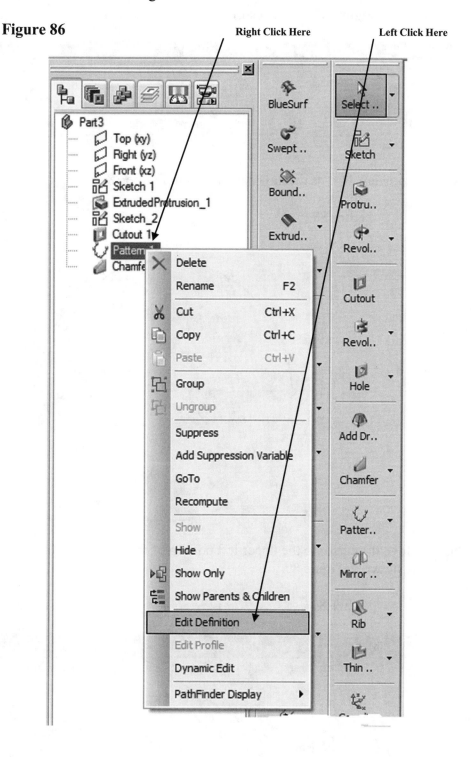

93. Left click on the **Select Curve Step** icon at the far left. Enter **6** for the number of holes and left click on the green checkmark as shown in Figure 87.

Figure 87   Left Click Here   Enter 6 Here   Left Click Here

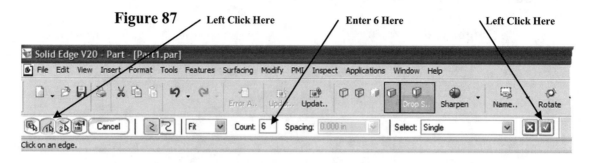

94. A preview will be displayed as shown in Figure 88.

Figure 88

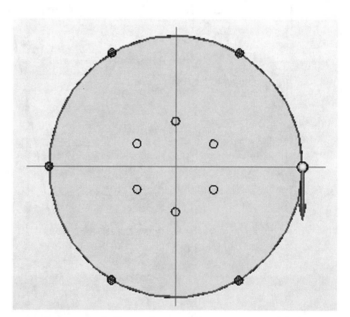

95. Move the cursor to the upper left portion of the screen and left click on **Preview** as shown in Figure 89.

Figure 89   Left Click Here

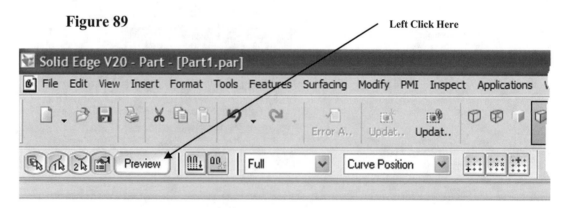

96. Move the cursor to the upper left portion of the screen and left click on **Finish** as shown in Figure 90.

Figure 90

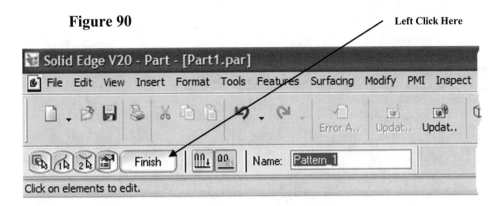

97. Your screen should look similar to Figure 91.

Figure 91

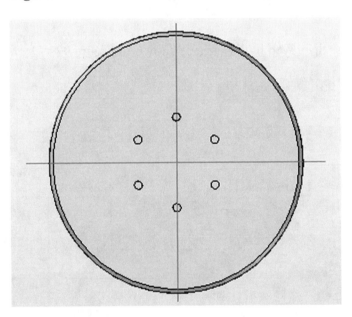

98. Move the cursor to the upper right portion of the screen and left click on the drop down arrow next to the "Named Views" icon. A drop down menu will appear. Left click on **dimetric** as shown in Figure 92.

Figure 92

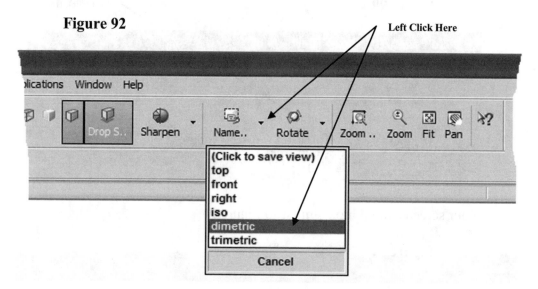

99. The part will be displayed in dimetric as shown in Figure 93.

Figure 93

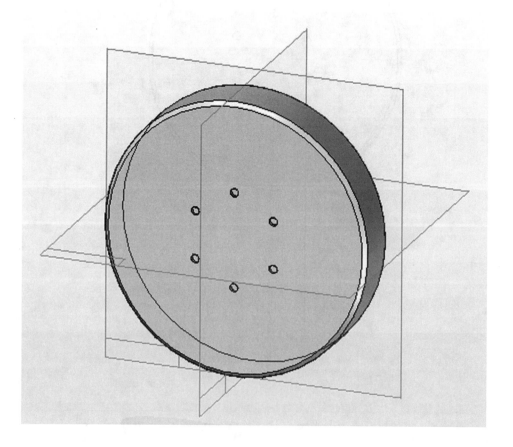

100. Move the cursor to the lower left portion of the screen where the part tree is located. Right click on **Chamfer 1**. A pop up menu will appear. Left click on **Edit Definition** as shown in Figure 94.

Figure 94

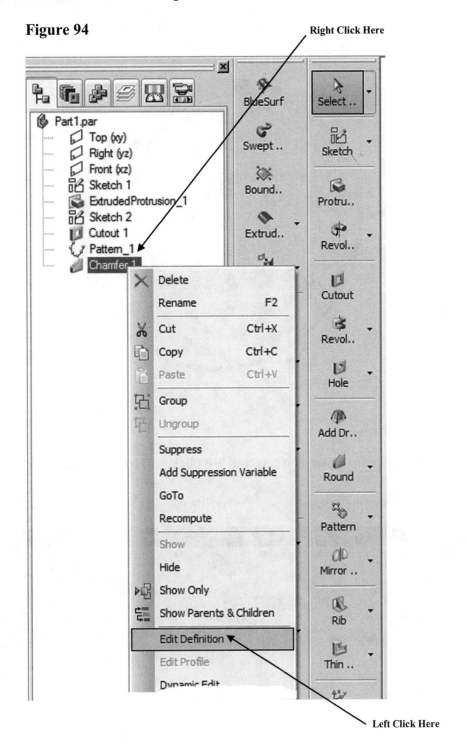

101. Move the cursor over the dimension and single left click as shown in Figure 95.

Figure 95

Single Left Click Here

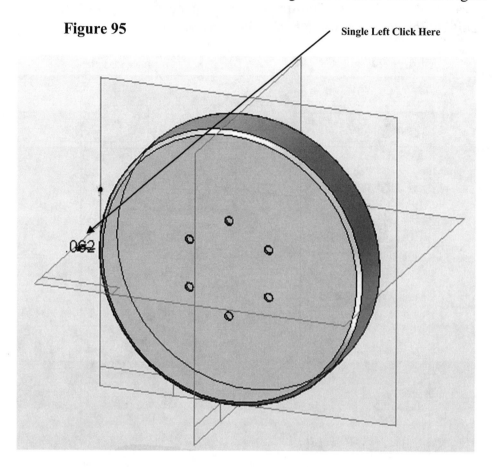

102. Enter **.25** in the dimension box and left click on **Finish** as shown in Figure 96.

Figure 96   Left Click Here   Enter .250 Here

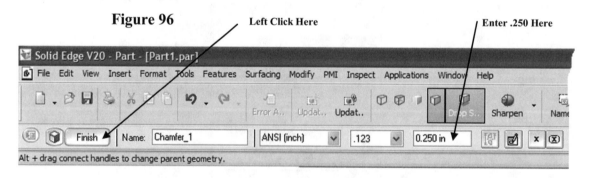

103. Your screen should look similar to Figure 97.

Figure 97

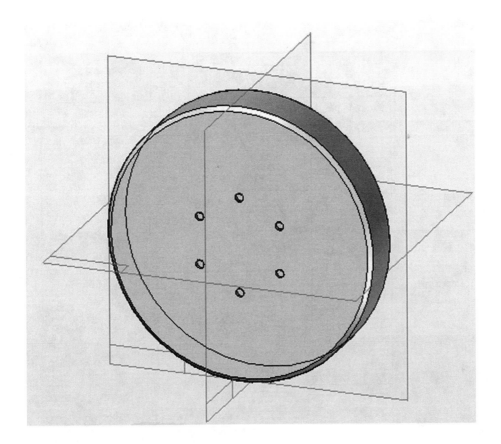

104. Move the cursor to the middle left portion of the screen where the part tree is located. Right click on **Pattern_1.** A pop up menu will appear. Left click on **Suppress** as shown in Figure 98.

Figure 98

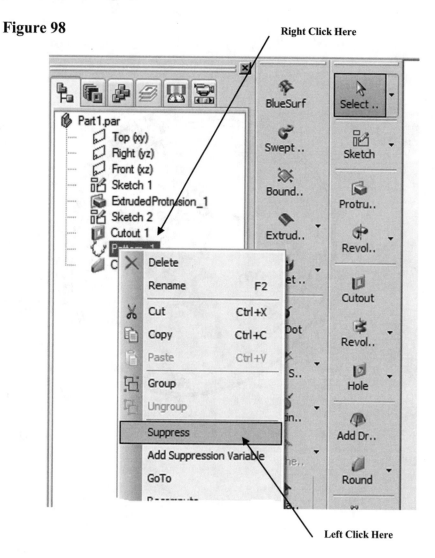

Right Click Here

Left Click Here

105. All holes created using the Pattern Along a Curve command will be suppressed except for the original hole as shown in Figure 99.

Figure 99

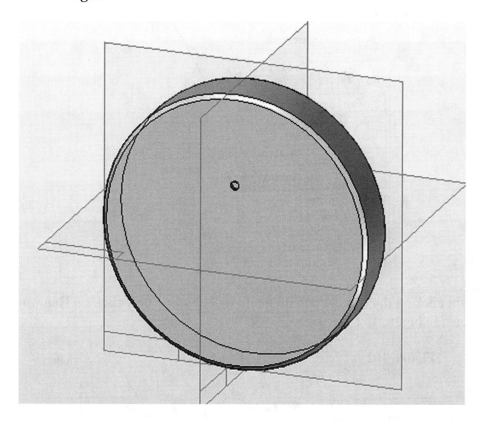

106. Solid Edge will place a small "no" symbol next to the suppressed feature as shown in Figure 100.

Figure 100

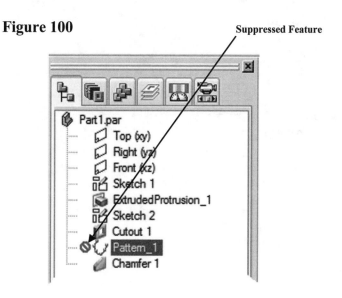

Suppressed Feature

107. The names of all branches in the part tree can also be edited. Left click once on the text **ExtrudedProtrusion_1**. The text will become highlighted as shown in Figure 101.

Figure 101

Highlighted Text

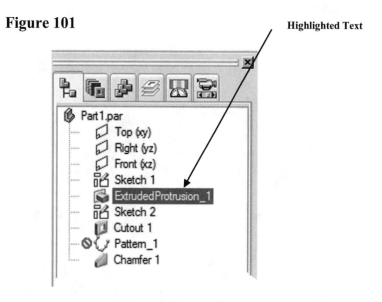

108. Right click on the text. A pop up menu will appear. Left click on **Rename** as shown in Figure 102.

Figure 102

Right Click Here   Left Click Here

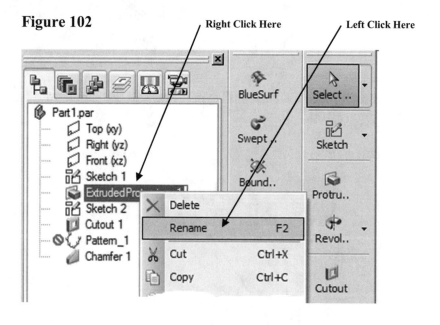

109. Enter the text **Base Protrusion** as shown in Figure 103. Press **Enter** on the keyboard. Text for each of the individual operations can be edited if desired.

Figure 103

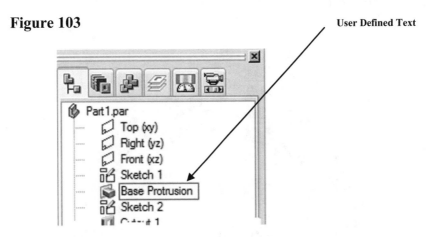

110. Move the cursor over the text **Pattern_1** and left click once. After the text becomes highlighted right click. A pop up menu will appear. Left click on **Unsuppress** as shown in Figure 104.

Figure 104

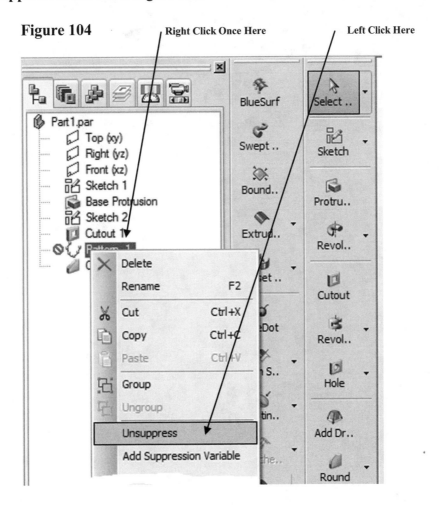

111. Your screen should look similar to Figure 105.

Figure 105

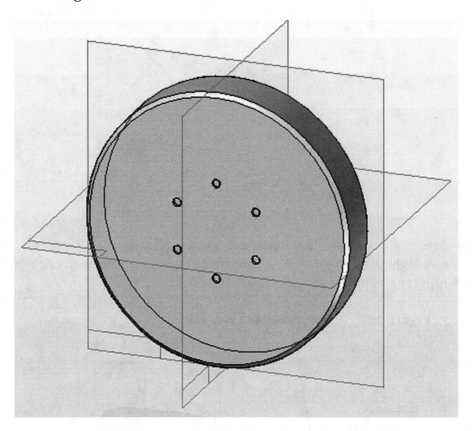

112. Notice that the final design looks significantly different than the original design. The new part was redesigned by modifying the existing part as shown in Figure 106.

Figure 106

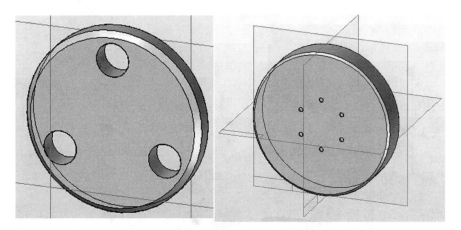

Drawing Activities

Use these problems from Chapters 1 and 2 to create redesigned parts.

Problem 1

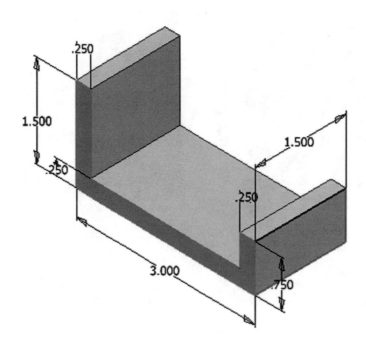

Problem 2

Extrude Center Section .25 Deep

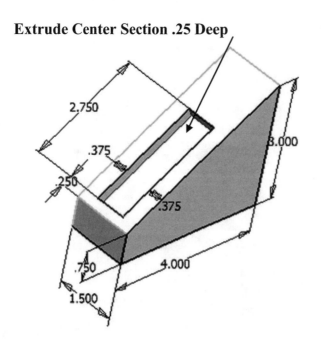

Problem 3

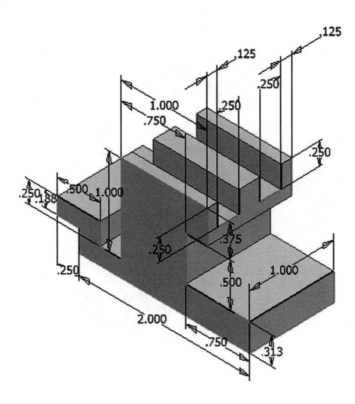

Problem 4

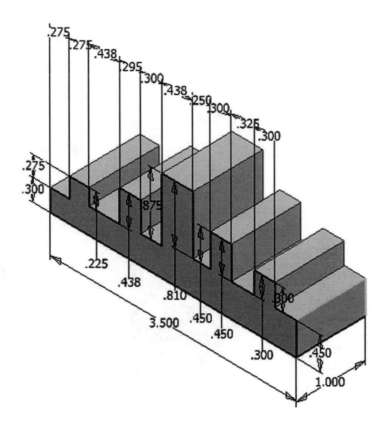

Problem 5

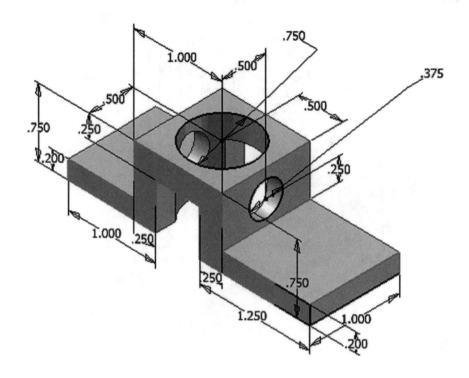

Problem 6

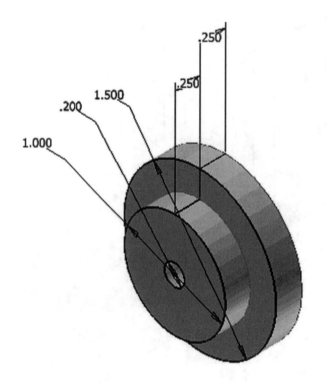

Problem 7

Revolve Axis

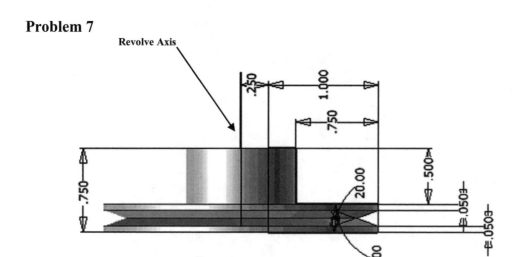

Problem 8

Revolve Axis

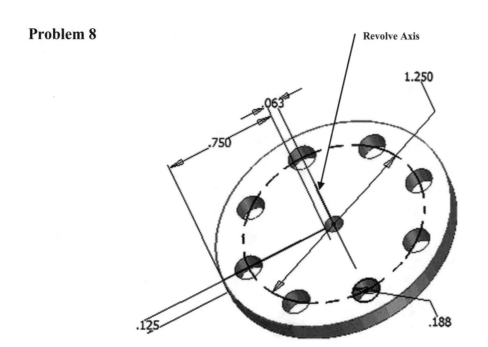

Chapter 6 Advanced Design Procedures

Objectives:
- Design multiple sketch parts
- Learn to use the Front, Top, and Right Planes
- Learn to use the Thin Wall command
- Learn to use the Shaded with Visible Edges command

Chapter 6 includes instruction on how to design the parts shown below.

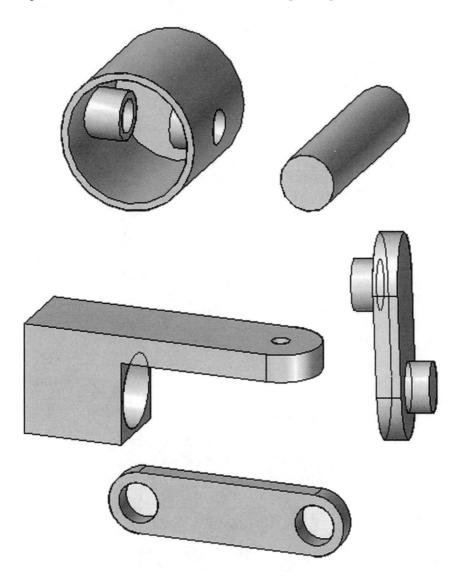

1.   Start Solid Edge by referring to "Chapter 1 Getting Started".

2.   After Solid Edge is running, begin a new sketch.

3.   Move the cursor to the upper middle portion of the screen and left click on **Circle** as shown in Figure 1.

Figure 1   Left Click Here

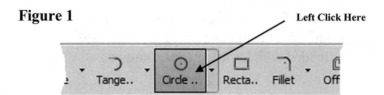

4.   Move the cursor to the center of the screen and left click once. This will be the center of the circle as shown in Figure 2.

Figure 2   Center of Circle

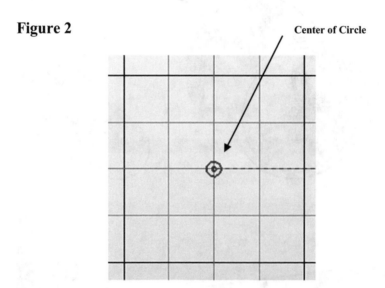

5.   Move the cursor to the right and left click once as shown in Figure 3.

Figure 3   Left Click Here

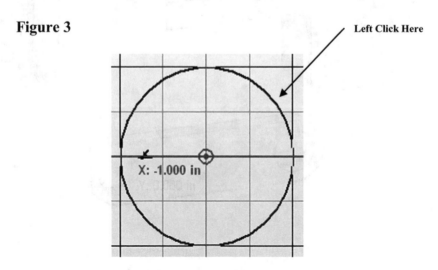

X: -1.000 in

6. Move the cursor to the upper left portion of the screen and left click on **SmartDimension** as shown in Figure 4.

Figure 4

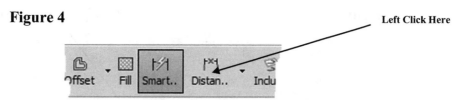

Left Click Here

7. After selecting **SmartDimension** move the cursor over the edge of the circle causing it to turn red as shown in Figure 5. Left click once. The dimension box will be attached to the cursor.

Figure 5

Turned Red

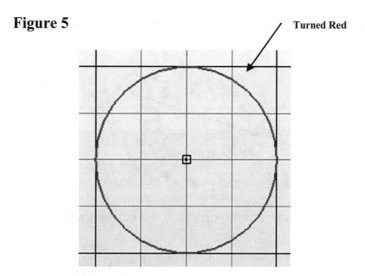

8. Move the cursor to where the dimension will be placed and left click once as shown in Figure 6.

Figure 6

Left Click Here

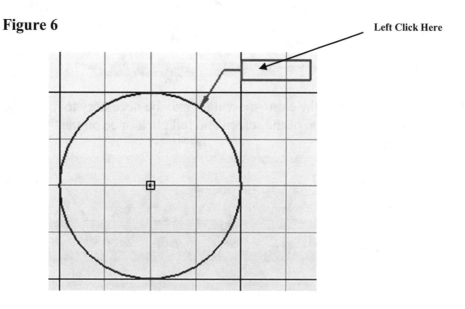

9. Move the cursor to the upper middle portion of the screen and enter **2.000** in the dimension box as shown in Figure 7.

Figure 7

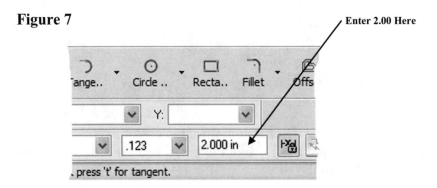

10. Move the cursor to the upper left portion of the screen and left click on **Return** as shown in Figure 8.

Figure 8

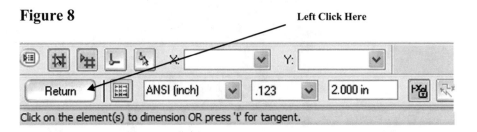

11. Move the cursor to the upper left portion of the screen and left click on **Finish** as shown in Figure 9.

Figure 9

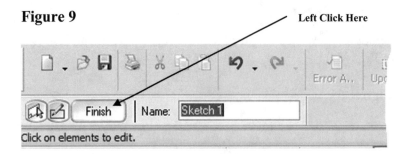

12. In order to view the entire drawing it may be necessary to move the cursor to the upper right portion of the screen and left click once on the "Fit" icon as shown in Figure 10.

Figure 10

13. The drawing will "fill up" the entire screen. If the drawing is still too large, left click on the "Zoom" icon as shown in Figure 11. After selecting the Zoom icon, hold the left mouse button down and drag the cursor up or down to achieve the desired view of the sketch.

Figure 11

14. Solid Edge is now out of the Sketch command and into the Model command. Notice that the commands at the top of the screen are now different. To work in the Model area a sketch must be present and have no opens (non-connected lines). If there are any opens in the sketch an error message will appear. Your screen should look similar to Figure 12.

Figure 12

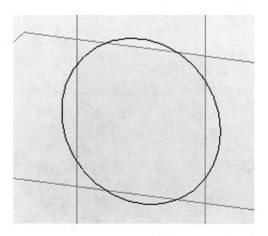

15. Move the cursor to the upper left portion of the screen and left click on **Protrusion** as shown in Figure 13.

Figure 13

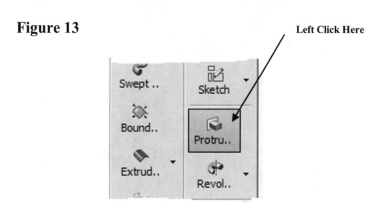

285

16. Move the cursor over the circle causing it to turn red. Left click then right click as shown in Figure 14.

Figure 14

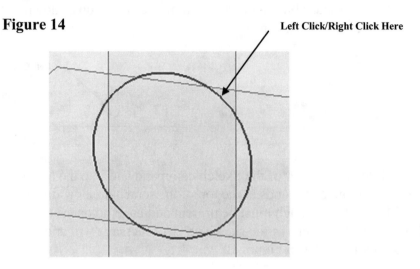

Left Click/Right Click Here

17. The sketch will become three dimensional as shown in Figure 15. The extruded surface will be attached to the cursor.

Figure 15

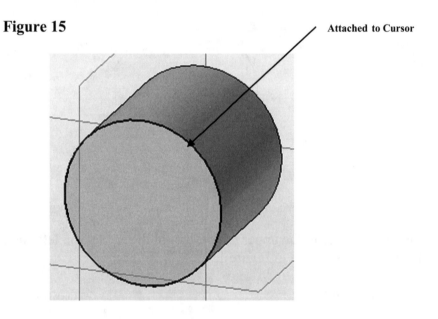

Attached to Cursor

18. Enter **2.00** for the Distance as shown in Figure 16.

Figure 16

Enter 2.00 Here

19. Move the cursor in the upper right direction of the screen and left click once. Your screen should look similar to Figure 17.

Figure 17

Left Click Here

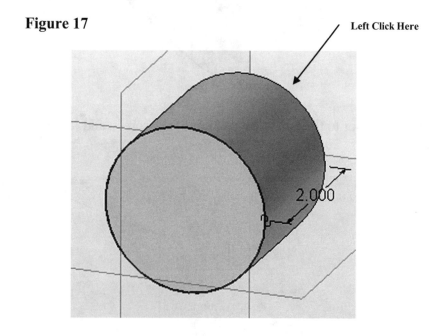

20. Move the cursor to the upper left portion of the screen and left click on **Finish** as shown in Figure 18.

Figure 18

Left Click Here

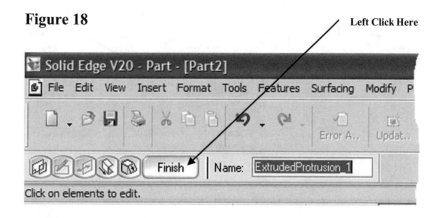

21.   Your screen should look similar to Figure 19. Press the **Esc** key once or twice.

Figure 19

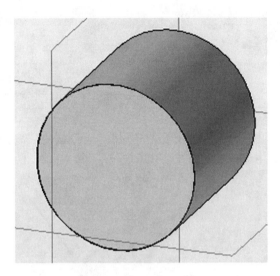

22.   Move the cursor to the upper left portion of the screen and left click on **Sketch** as shown in Figure 20.

Figure 20

Left Click Here

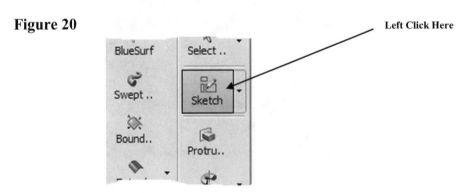

23.   Move the cursor to the upper left portion of the screen and left click on the text "Right {yz}" as shown in Figure 21.

Figure 21

Left Click Here

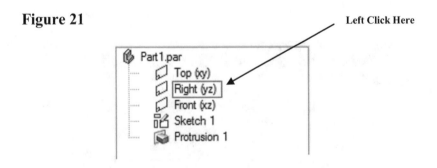

24. Solid Edge will rotate the model. Your screen should look similar to Figure 22.

Figure 22

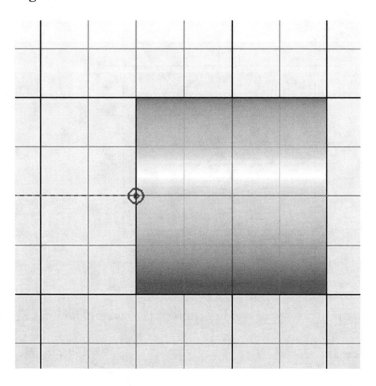

25. Move the cursor to the upper right portion of the screen and left click on the "Visible and Hidden Edges" icon as shown in Figure 23.

Figure 23

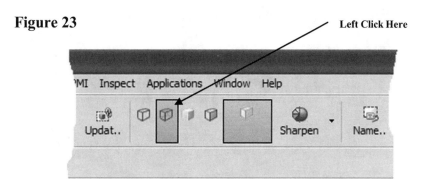

26. Solid Edge has hidden the model from view. Your screen should look similar to Figure 24.

Figure 24
Center of Model

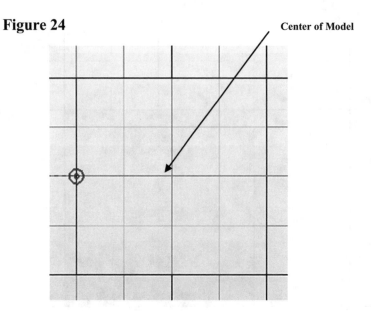

27. Move the cursor to the upper right portion of the screen and left click on **Include**.

Figure 25
Left Click Here

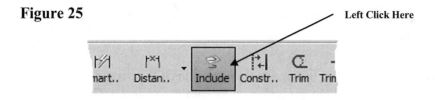

28. The Include Options dialog box will appear. Left click on **OK**.

Figure 26
Left Click Here

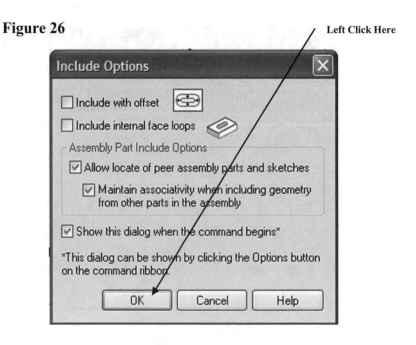

29. The lines that make up the part may not be visible. Move the cursor around where they were located. The lines will appear in red. Move the cursor over the top line causing it to turn red and left click once as shown in Figure 27.

Figure 27

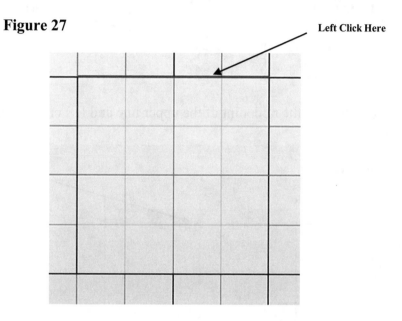

30. Move the cursor over the lower line causing it to turn red. Left click once as shown in Figure 28.

Figure 28

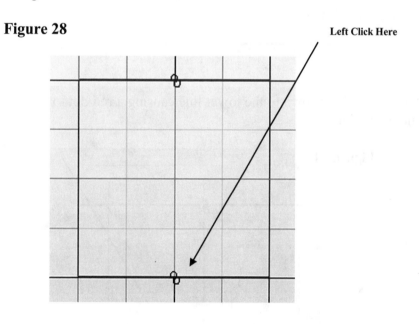

31. Move the cursor to the upper left portion of the screen and left click on **Line** as shown in Figure 29.

Figure 29

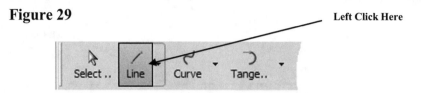

32. Move the cursor to the midpoint of the upper line and left click as shown in Figure 30.

Figure 30

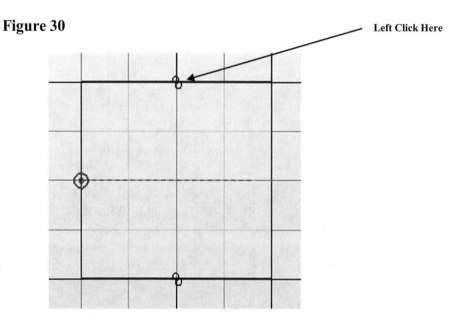

33. Move the cursor down to the lower line causing a red dot to appear. Left click as shown in Figure 31.

Figure 31

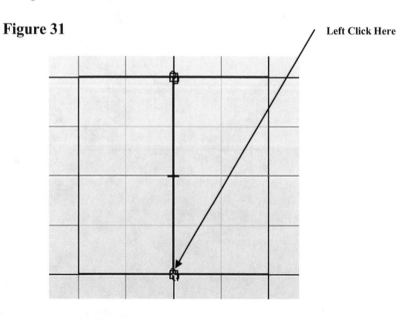

34. Move the cursor to the upper middle portion of the screen and left click on **Circle** as shown in Figure 32.

Figure 32

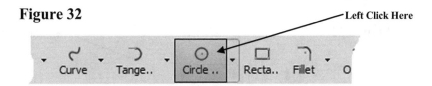

35. Move the cursor to the midpoint of the center line and left click as shown in Figure 33.

Figure 33

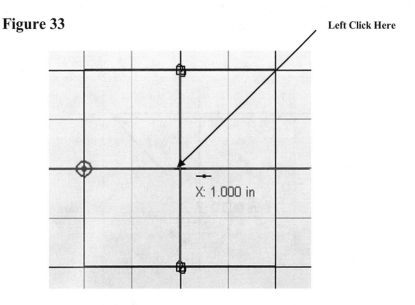

36. Move the cursor out to the side and left click as shown in Figure 34.

Figure 34

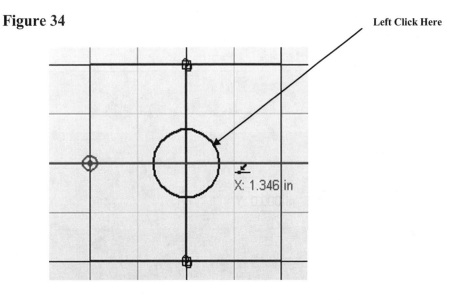

37. Move the cursor to the upper left portion of the screen and left click on **SmartDimension** as shown in Figure 35.

Figure 35

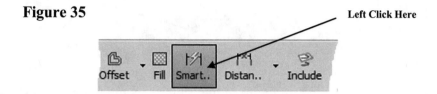

Left Click Here

38. Move the cursor over the edge (not center) of the circle until it turns red. Left click once as shown in Figure 36.

Figure 36

Left Click Here

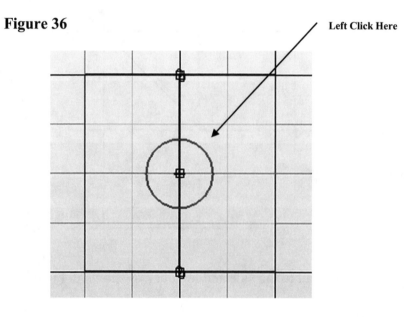

39. The dimension box will be attached to the cursor. Left click as shown in Figure 37.

Figure 37

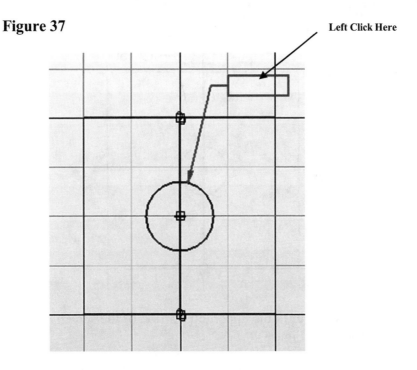

Left Click Here

40. Enter **.500** in the dimension box as shown in Figure 38. Press the **Enter** key.

Figure 38

Enter .500 Here

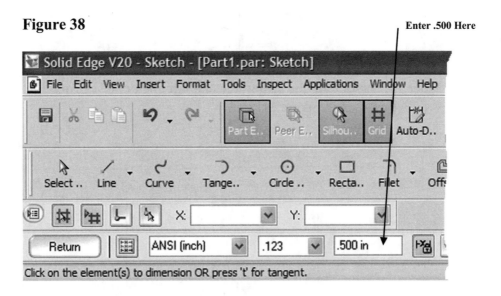

41. Your screen should look similar to Figure 39.

Figure 39

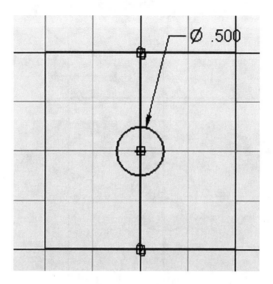

42. Move the cursor over to the upper left portion of the screen and left click on **Select** as shown in Figure 40.

Figure 40

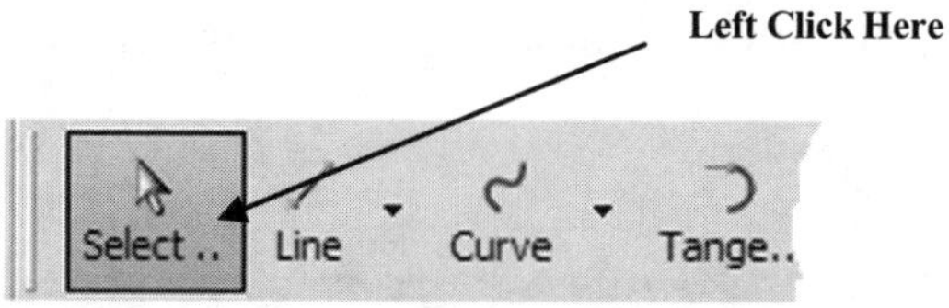

Left Click Here

43. Move the cursor over the center line and left click once causing the line to turn yellow as shown in Figure 41.

Figure 41

Left Click Here

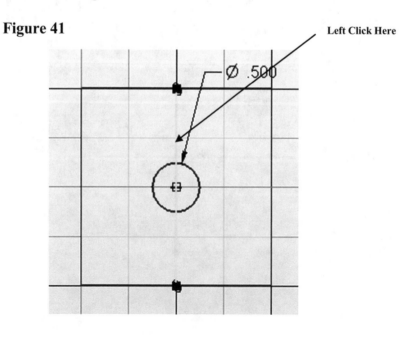

44. Move the cursor to the upper left portion of the screen and left click on **Edit**. A drop down menu will appear. Left click on **Delete** as shown in Figure 42.

Figure 42

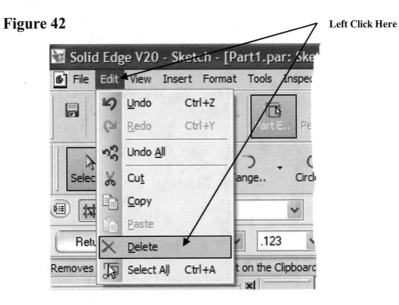

45. Repeat the same steps to delete the other lines projected on to the Right Plane. Your screen should look similar to Figure 43.

Figure 43

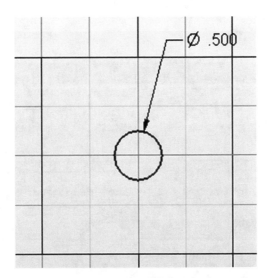

46. Move the cursor to the upper right portion of the screen and left click on the "Shaded with Visible Edges" icon as shown in Figure 44.

Figure 44

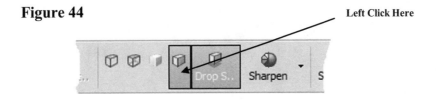

47. Your screen should look similar to Figure 45.

Figure 45

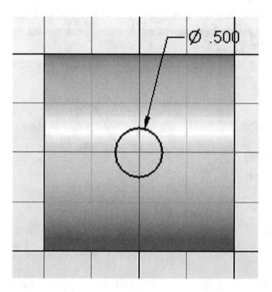

48. After you have verified that no commands are active, move the cursor to the upper left portion of the screen and left click on **Return** as shown in Figure 46.

Figure 46

Left Click Here

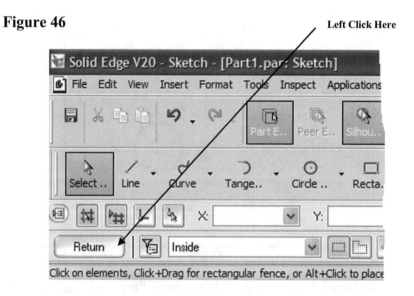

49. Move the cursor to the upper left portion of the screen and left click on **Finish** as shown in Figure 47.

Figure 47

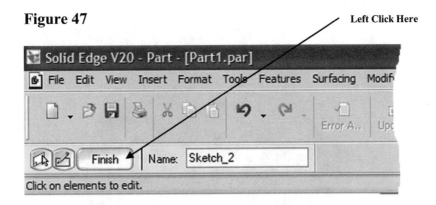

Left Click Here

50. Move the cursor to the upper right portion of the screen and left click on the drop down arrow next to the "Named Views" icon. A drop down menu will appear. Left click on **trimetric** as shown in Figure 48.

Figure 48

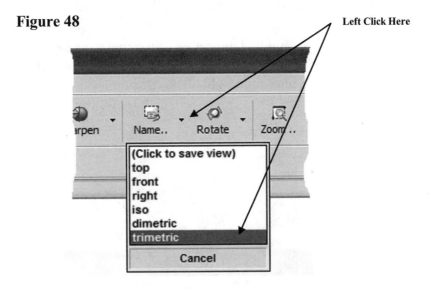

Left Click Here

51. Your screen should look similar to Figure 49.

Figure 49

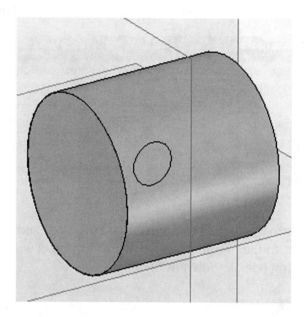

52. Move the cursor to the middle left portion of the screen and left click on **Cutout** as shown in Figure 50. Move the cursor to the edge of the small circle causing it to turn red. Left click once then right click.

Figure 50

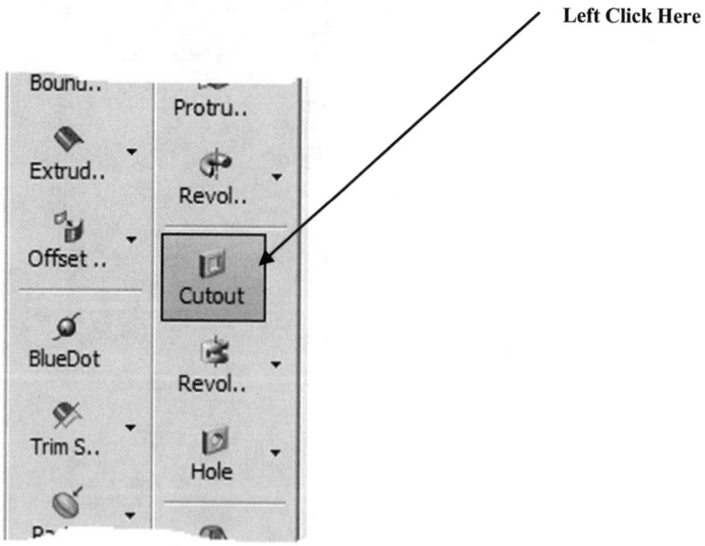

Left Click Here

300

53. Move the cursor to the upper left portion of the screen and left click on the "Symmetric Extent" icon as shown in Figure 61. The cutout will be attached to the cursor.

Figure 51

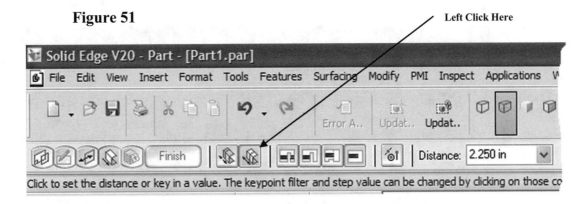

54. A preview will be displayed as shown in Figure 52.

Figure 52

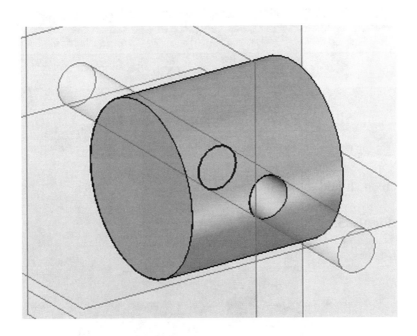

301

55. Move the cursor to the upper middle portion of the screen and enter **2.250** for the Distance as shown in Figure 53. Press the **Enter** key on the keyboard.

Figure 53

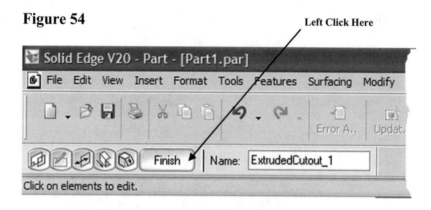

56. Move the cursor to the upper left portion of the screen and left click on **Finish** as shown in Figure 54.

Figure 54

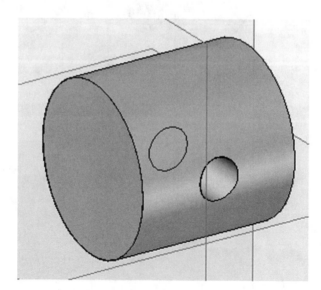

57. Your screen should look similar to Figure 55.

Figure 55

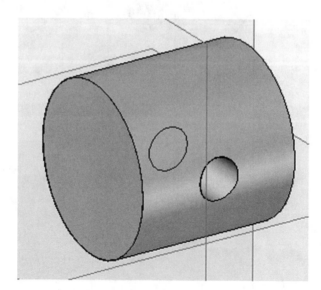

58. Move the cursor to the lower right portion of the screen and left click on **Thin Wall** as shown in Figure 56.

Figure 56

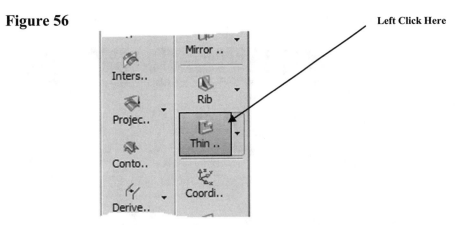

Left Click Here

59. Enter **.100** for Common thickness as shown in Figure 57. Press the **Enter** key on the keyboard.

Figure 57

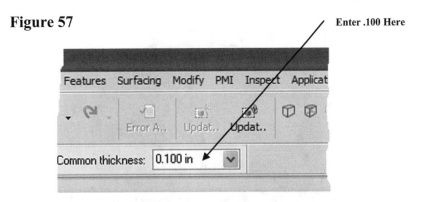

Enter .100 Here

60. Move the cursor to the lower surface causing it to turn red. Left click then right click as shown in Figure 58.

Figure 58

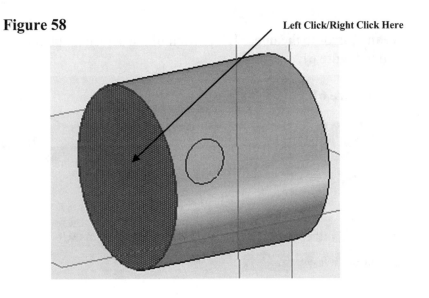

Left Click/Right Click Here

61.   Move the cursor to the upper left portion of the screen and left click on **Preview** as shown in Figure 59.

Figure 59

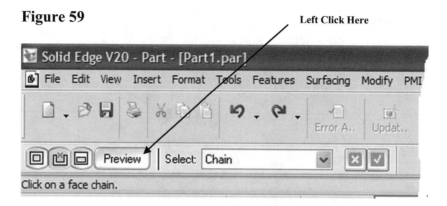

62.   Use the Rotate command to roll the part around to see the underside. Your screen should look similar to Figure 60.

Figure 60

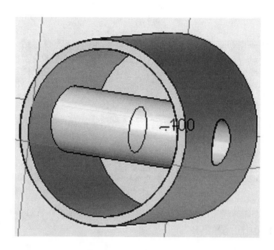

63.   Move the cursor to the upper left portion of the screen and left click on **Finish** as shown in Figure 61.

Figure 61

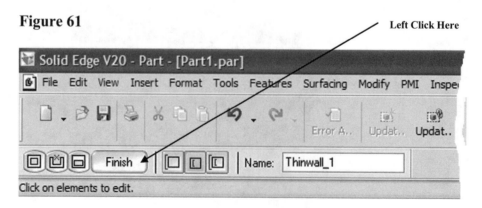

64. Move the cursor to the upper right portion of the screen and left click on the drop down arrow next to the "Named Views" icon. A drop down menu will appear. Left click on **front** as shown in Figure 62.

Figure 62

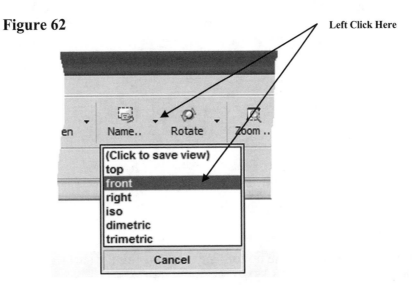

65. The part will rotate to provide a perpendicular view of the inside as shown in Figure 63.

Figure 63

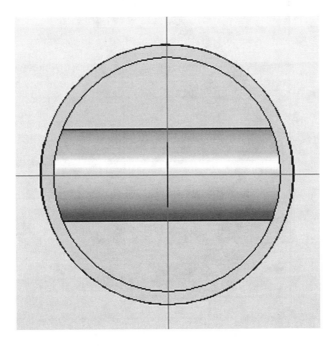

66. Move the cursor to the upper middle portion of the screen and left click on **Sketch** as shown in Figure 64.

Figure 64

Left Click Here

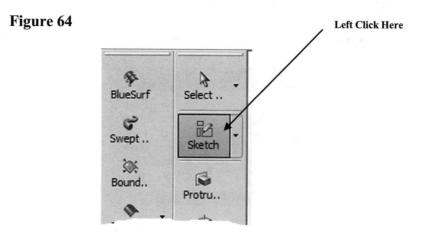

67. Move the cursor over the bottom surface causing the surface to turn red and left click as shown in Figure 65.

Figure 65

Left Click Here

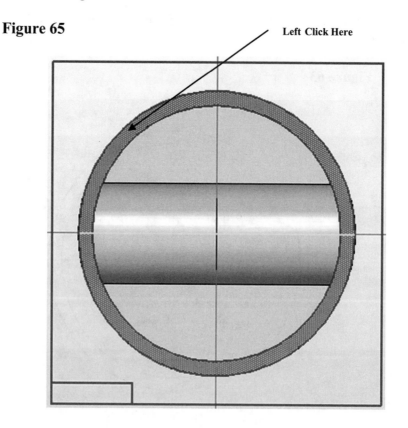

68. A new sketch will appear on the selected surface. Your screen should look similar to Figure 66.

Figure 66

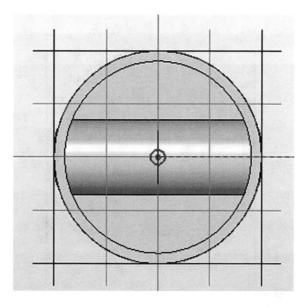

69. Move the cursor to the upper left portion of the screen and left click on **Line** as shown in Figure 67.

Figure 67

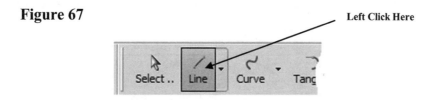

70. Move the cursor to the center of the part and left click on the red dot as shown in Figure 68.

Figure 68

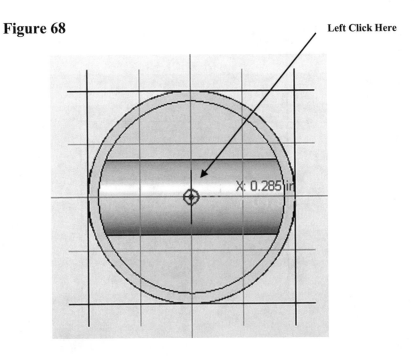

71. Move the cursor upward and left click as shown in Figure 69.

Figure 69

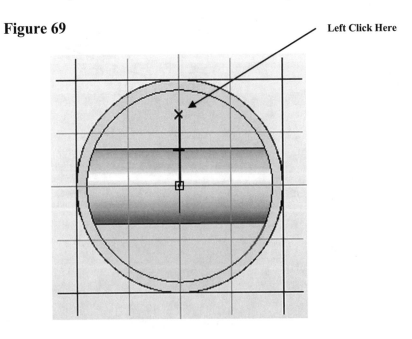

72. Move the cursor to the upper left portion of the screen and left click on **Line** as shown in Figure 70.

Figure 70

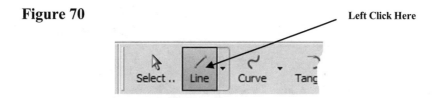

Left Click Here

73. Move the cursor to the center of the part and left click on the red dot as shown in Figure 71.

Figure 71

Left Click Here

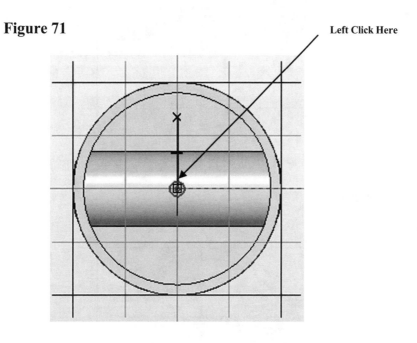

74. Move the cursor to the left and left click as shown in Figure 72.

Figure 72

Left Click Here

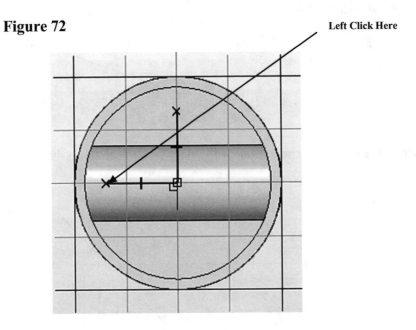

75. Your screen should look similar to Figure 73.

Figure 73

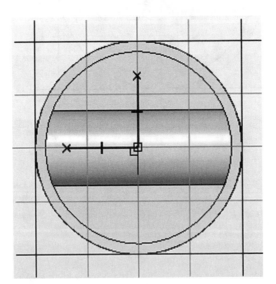

76. Move the cursor to the upper left portion of the screen and left click on **Rectangle** as shown in Figure 74.

Figure 74

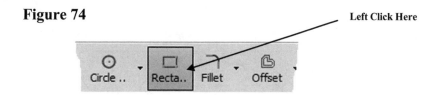

Left Click Here

77. Move the cursor to the position shown in Figure 75 and left click once.

Figure 75

Left Click Here

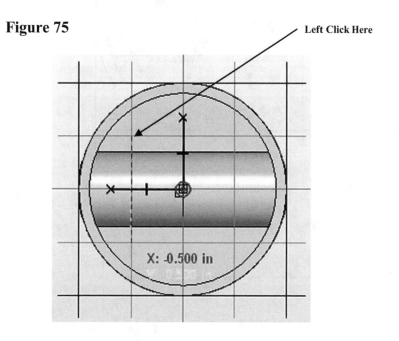

X: -0.500 in

78. Move the cursor downward and left click as shown in Figure 76.

Figure 76

Left Click Here

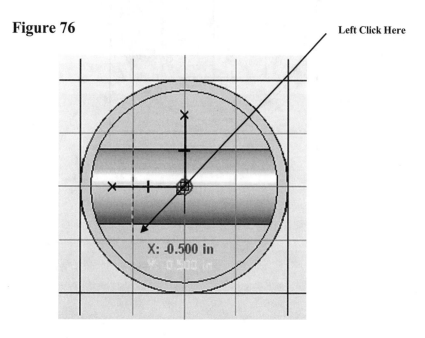

X: -0.500 in

311

79. Move the cursor to the right and left click as shown in Figure 77.

Figure 77

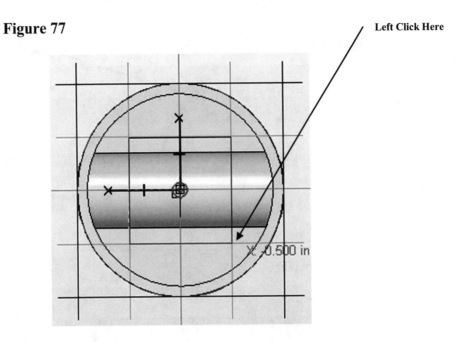

Left Click Here

80. Your screen should look similar to Figure 78.

Figure 78

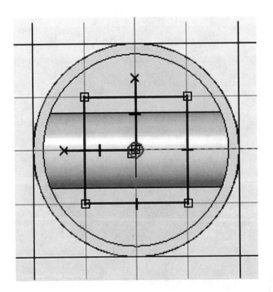

81. Move the cursor to the middle left portion of the screen and left click on **SmartDimension** as shown in Figure 79.

Figure 79

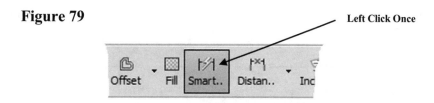

Left Click Once

82. After selecting **SmartDimension** move the cursor over the vertical line coming out of the center of the part causing it to turn red. Left click as shown in Figure 80.

Figure 80

Left Click Here

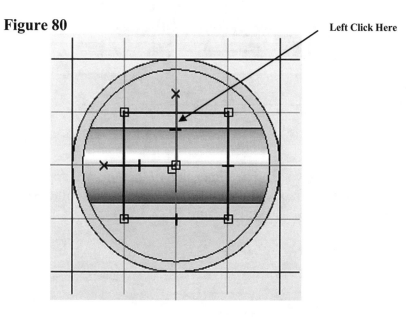

83. Move the cursor to the far left line and left click once as shown in Figure 81. Ignore the dimension box that appears.

Figure 81

Left Click Here

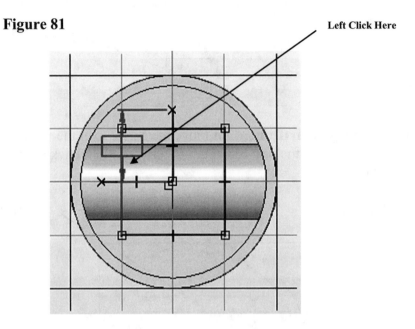

84. Move the cursor upward. The dimension of the line will appear. The dimension is attached to the cursor. Move the cursor to where the dimension will be placed and left click once as shown in Figure 82.

Figure 82

Left Click Here

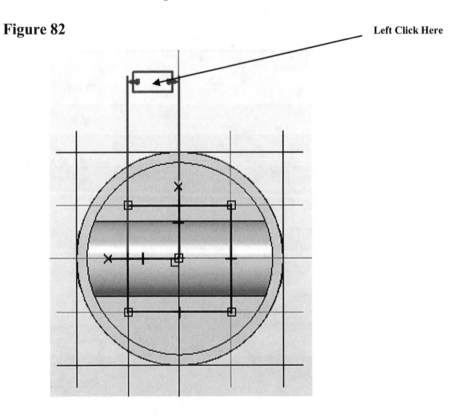

85. Enter **.375** in the dimension box as shown in Figure 83. Press the **Enter** key on the keyboard.

Figure 83   Enter .375 Here

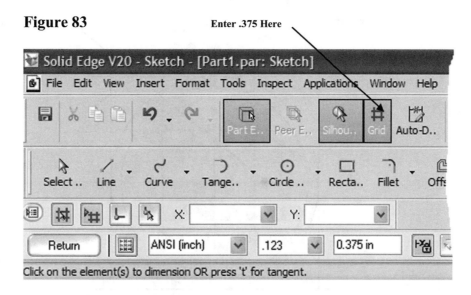

86. The dimension of the line will become .375 inches as shown in Figure 84.

Figure 84

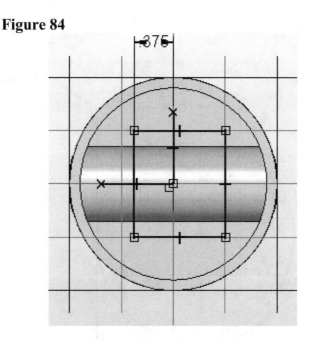

87. Move the cursor to the middle left portion of the screen and left click on **SmartDimension** as shown in Figure 85.

Figure 85   Left Click Here

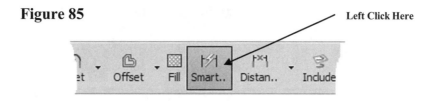

88.   Move the cursor over the horizontal line coming out of the center causing it to turn red. Left click as shown in Figure 86.

Figure 86

Left Click Here

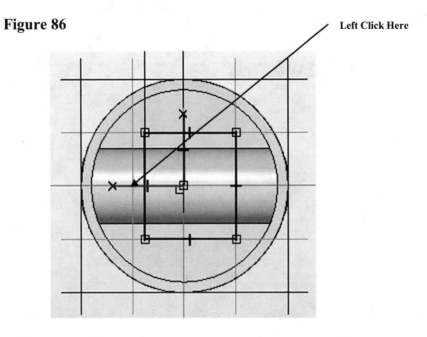

89.   Move the cursor to the upper line and left click once as shown in Figure 87. Ignore the dimension box.

Figure 87

Left Click Here

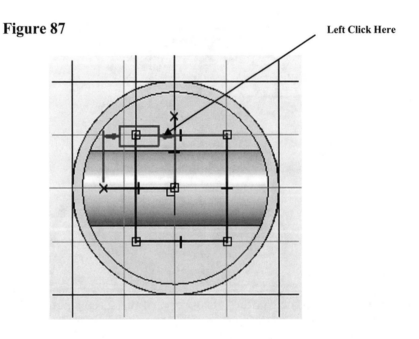

90. Move the cursor out to the side. The dimension of the line will appear and is attached to the cursor. Move the cursor to where the dimension will be placed and left click once as shown in Figure 88.

Figure 88

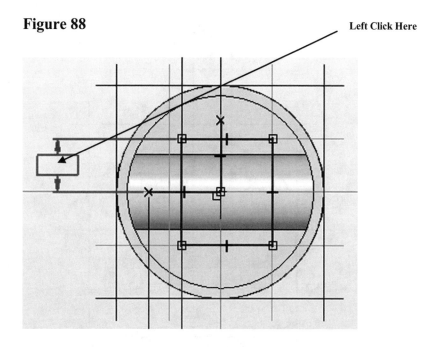

91. Enter **.500** in the dimension box as shown in Figure 89. Press the **Enter** key on the keyboard.

Figure 89

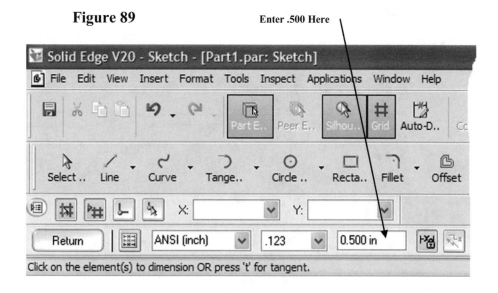

92. The dimension of the line will become .500 inches as shown in Figure 90.

Figure 90

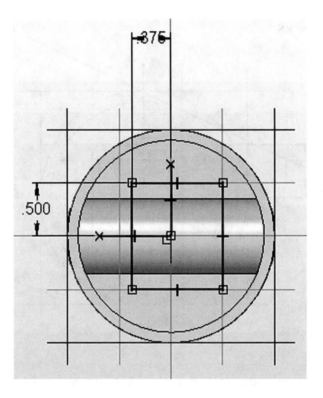

93. Move the cursor to the middle left portion of the screen and left click on **SmartDimension** as shown in Figure 91.

Figure 91

Left Click Here

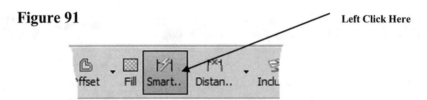

94. Move the cursor over the horizontal line coming out of the center of the part causing it to turn red. Left click as shown in Figure 92.

Figure 92

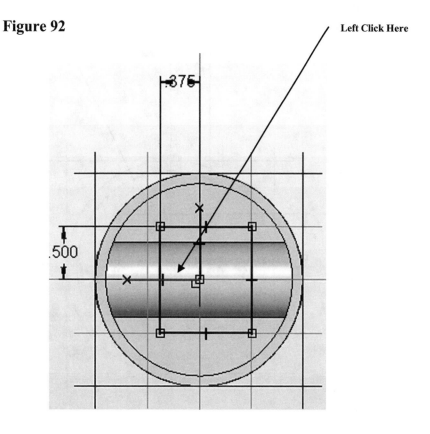

95.   Move the cursor to the lower horizontal line causing it to turn red. Left click as shown in Figure 93. Ignore the dimension box.

Figure 93

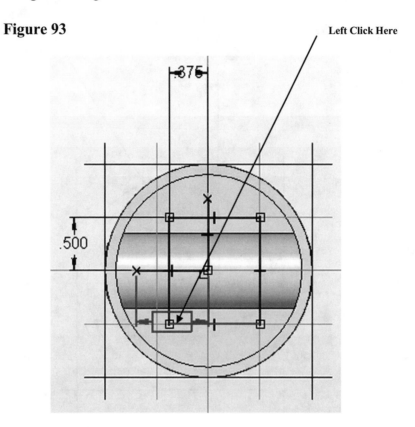

96. Move the cursor out to the side. The dimension of the line will appear and is attached to the cursor. Move the cursor to where the dimension will be placed and left click once as shown in Figure 94.

Figure 94

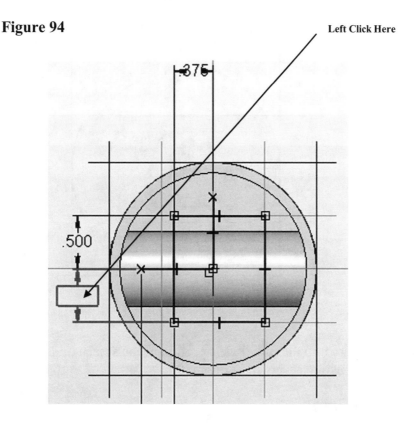

Left Click Here

97. Enter **.500** in the dimension box as shown in Figure 95. Press the **Enter** key on the keyboard.

Figure 95

Enter .500 Here

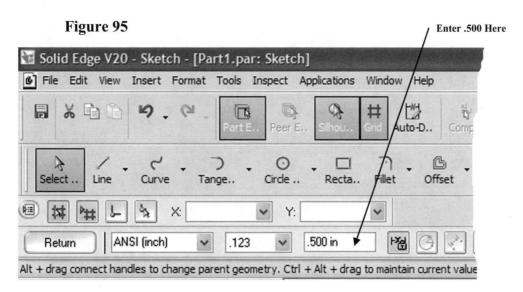

98. The dimension of the line will become .500 inches as shown in Figure 96.

Figure 96

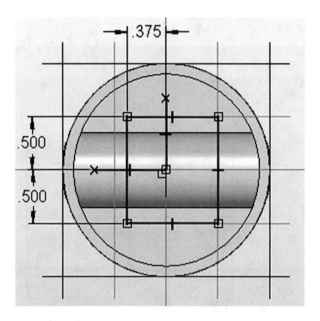

99. Move the cursor to the middle left portion of the screen and left click on **SmartDimension** as shown in Figure 97.

Figure 97   Left Click Here

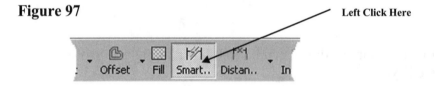

100. After selecting **SmartDimension** move the cursor to the vertical line coming out of the center of the part causing it to turn red. Left click as shown in Figure 98.

Figure 98

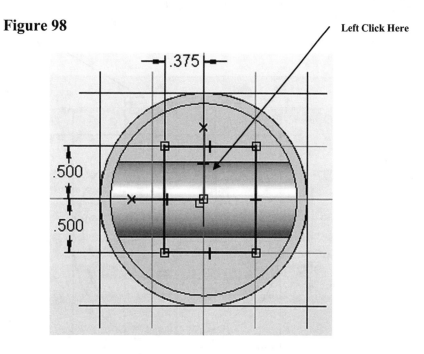

Left Click Here

101. Move the cursor to the right side vertical line causing it to turn red and left click once as shown in Figure 99. Ignore the dimension box.

Figure 99

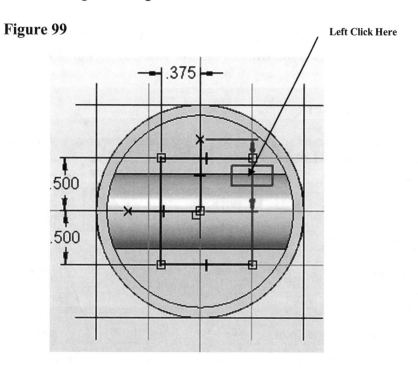

Left Click Here

102. Move the cursor out to the side. The dimension of the line will appear and is attached to the cursor. Move the cursor to where the dimension will be placed and left click once as shown in Figure 100.

Figure 100

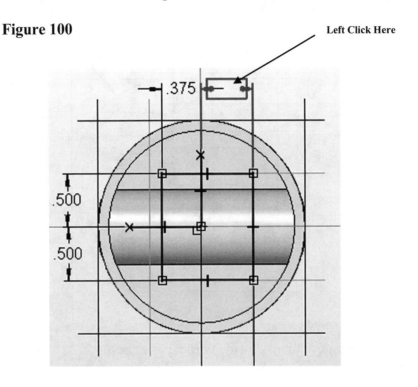

103. Enter **.375** in the dimension box as shown in Figure 101. Press the **Enter** key on the keyboard. Press the **Esc** key once or twice to get out of the SmartDimension command.

Figure 101

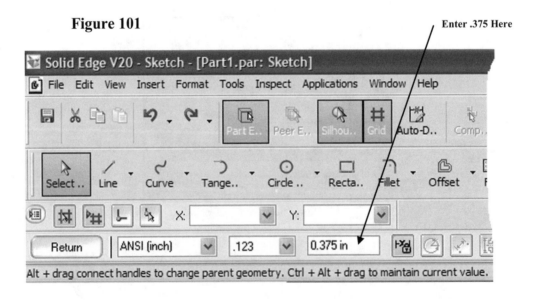

104. The dimension of the line will become .375 inches as shown in Figure 102.

Figure 102

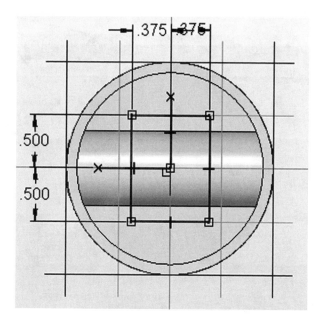

105. Move the cursor over the vertical line coming out of the center of the part causing it to turn red. Left click once causing it to turn yellow as shown in Figure 103.

Figure 103

Turned Red

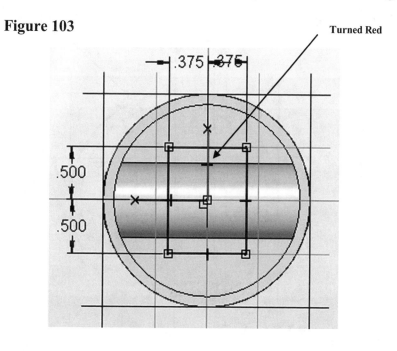

106. Move the cursor to the upper left portion of the screen and left click on **Edit**. A drop down menu will appear. Left click on **Delete** as shown in Figure 104.

Figure 104

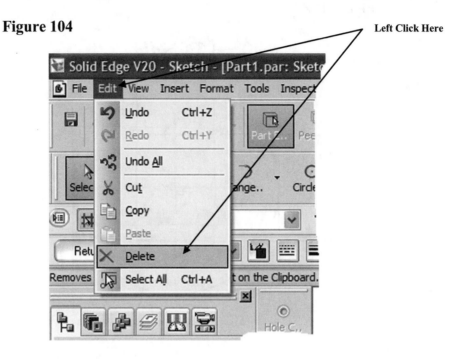

107. Use the same steps to delete the horizontal line coming out of the center of the part as shown in Figure 105. If the dimensions disappear do not be concerned.

Figure 105

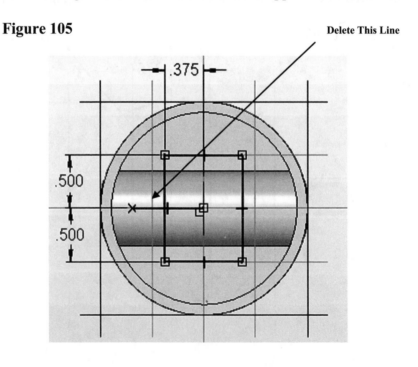

108. Move the cursor to the upper left portion of the screen and left click on **Return** as shown in Figure 106.

Figure 106

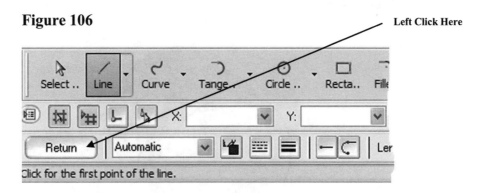

Left Click Here

109. Move the cursor to the upper left portion of the screen and left click on **Finish** as shown in Figure 107.

Figure 107

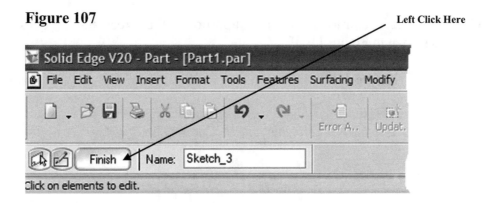

Left Click Here

110. Move the cursor to the upper right portion of the screen and left click on the drop down arrow next to the "Named Views" icon. A drop down menu will appear. Left click on **dimetric** as shown in Figure 108.

Figure 108

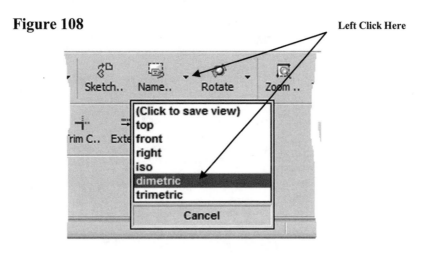

Left Click Here

111.   Your screen should look similar to Figure 109.

Figure 109

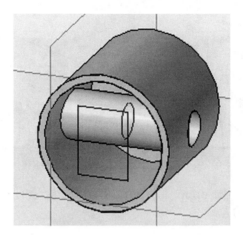

112.   Move the cursor to the middle left portion of the screen and left click on **Cutout** as shown in Figure 110.

Figure 110

Left Click Here

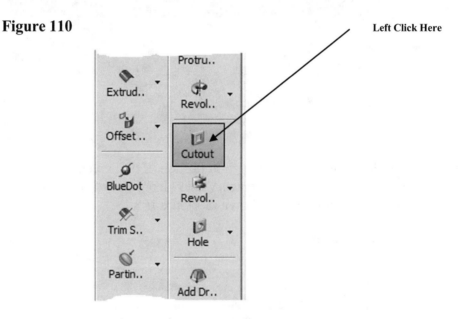

113.   Move the cursor to the box that was just drawn. Left click then right click on any of the lines as shown in Figure 111. The cutout will be attached to the cursor.

Figure 111   **Left Click/Right Click Here**

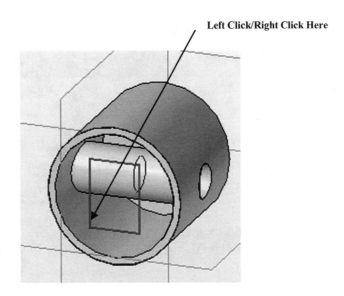

114.   Enter **1.875** for the Distance as shown in Figure 112.

Figure 112   **Enter 1.875 Here**

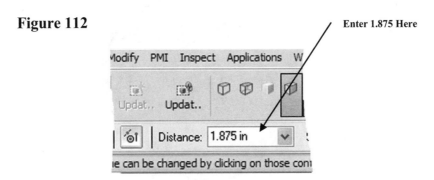

115.   Move the cursor to the back of the part and left click as shown in Figure 113.

Figure 113   **Left Click Here**

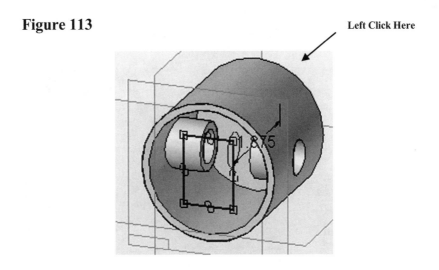

116. Move the cursor to the upper left portion of the screen and left click on **Finish** as shown in Figure 114.

Figure 114

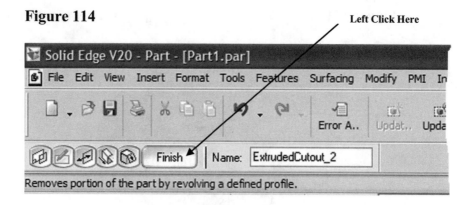

117. Your screen should look similar to Figure 115. Use the Rotate command to roll the part around to view the inside. Press the **Esc** key once or twice.

Figure 115

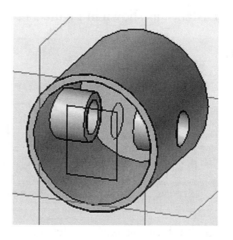

118. Move the cursor to the left portion of the screen towards the part tree and single left click on **Sketch_2**. Right click once. A drop down menu will appear. Left click on **Hide** as shown in Figure 116.

Figure 116

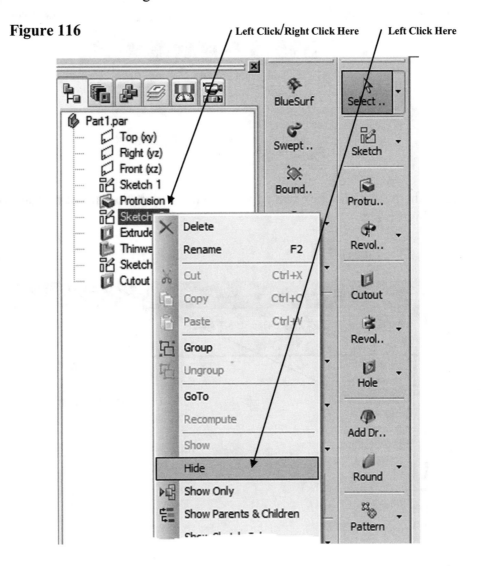

119. Follow the same steps to Hide Sketch_3.

120. Save the part as **Piston1.par** where it can be easily retrieved.

121. Move the cursor to the upper left portion of the screen and left click on the "New" icon as shown in Figure 117.

Figure 117

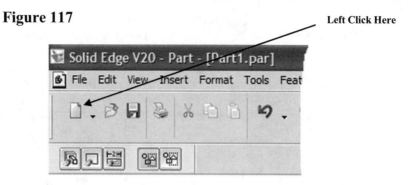

Left Click Here

122. The New dialog box will appear. Left click on **Normal.par** as shown in Figure 118.

Figure 118

Left Click Here

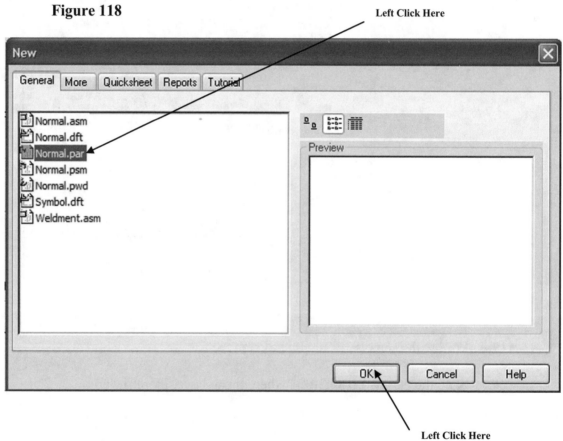

Left Click Here

123. Left click on **OK**.

124. Draw a circle in the center of the grid as shown in Figure 119.

Figure 119

Left Click Here

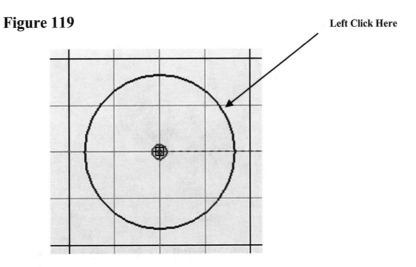

125. Use the **Smart Dimension** command to dimension the circle to **.500** inches as shown in Figure 120.

Figure 120

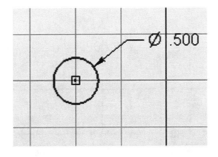

126. Exit the Sketch command and create a **Protrusion** (extrude) to a length of **1.875** inches as shown in Figure 121.

Figure 121

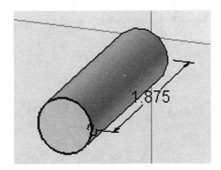

127. Your screen should look similar to Figure 122.

Figure 122

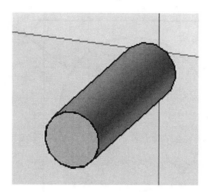

128. Save the part as **Wristpin1.par** where it can be easily retrieved later.

129. Begin a new sketch as previously described in this chapter.

130. Complete the sketch shown in Figure 123.

Figure 123

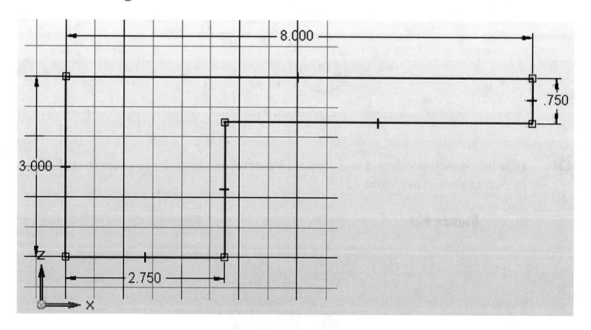

131. Exit the Sketch command. A trimetric view will be displayed as shown in Figure 124.

Figure 124

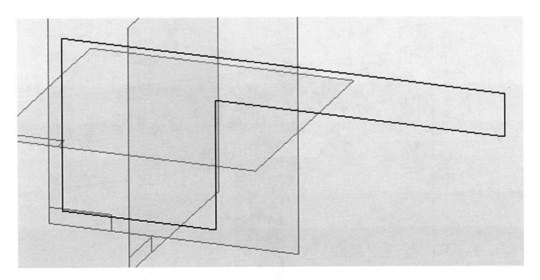

132. Create a protrusion (extrude) the sketch to a distance of **2.25** inches. Your screen should look similar to Figure 125.

Figure 125

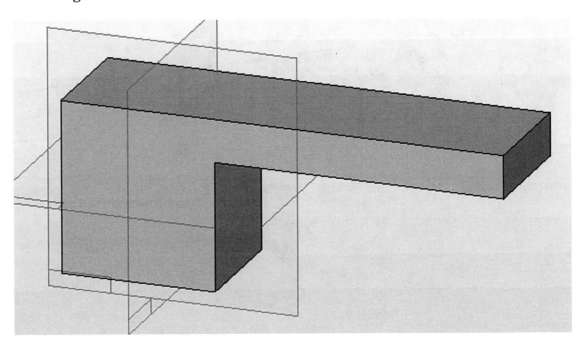

133. Use the **Round** command to create **1.125** inch rounds on the front portion of the part as shown in Figure 126.

Figure 126

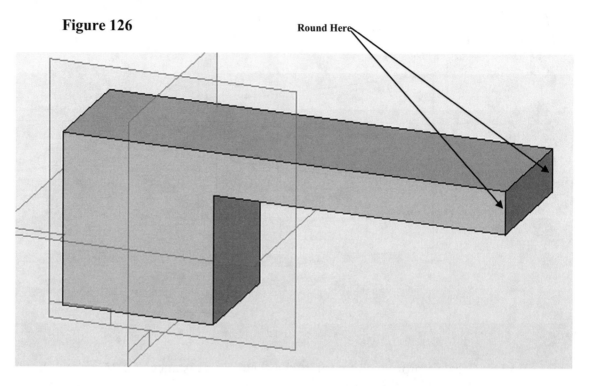

134. Your screen should look similar to Figure 127.

Figure 127

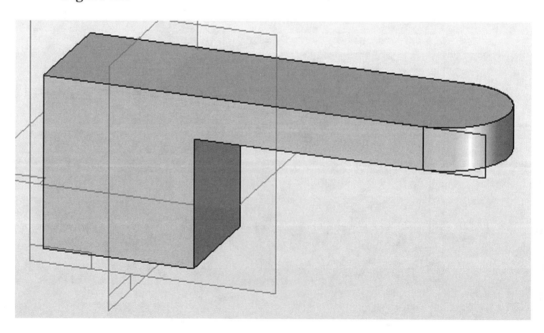

135. Use procedures described previously in this chapter to hide Sketch_1. Use these same procedures to hide the Top, Right and Front planes.

136. Your screen should look similar to Figure 128.

Figure 128

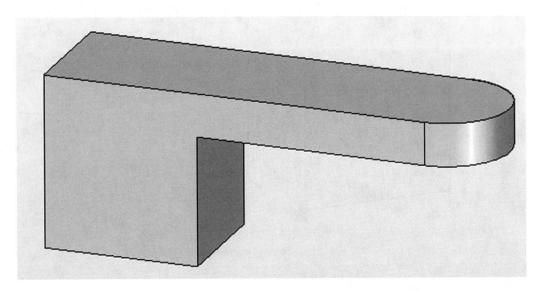

137. Move the cursor to the upper left portion of the screen and left click on **Sketch** as shown in Figure 129.

Figure 129

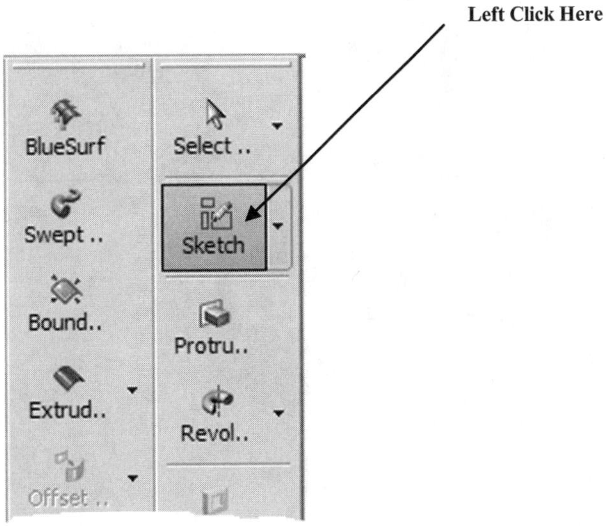

Left Click Here

138. Move the cursor to the surface shown in Figure 130 causing it to turn red.

Figure 130

Move Cursor Here

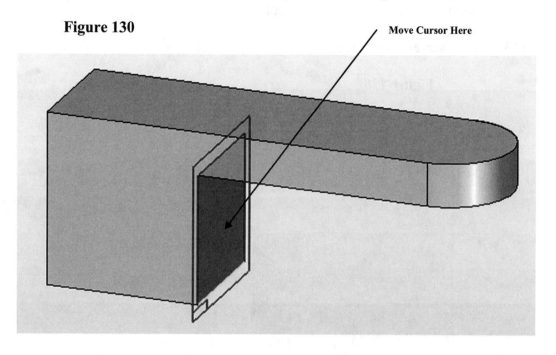

139. Left click once as shown in Figure 131.

Figure 131

Left Click Here

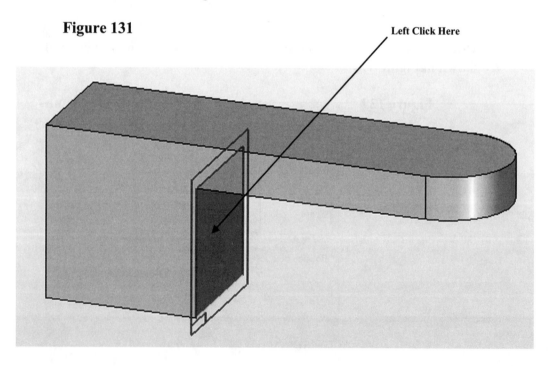

140. A perpendicular view of the surface will be displayed as shown in Figure 132.

Figure 132

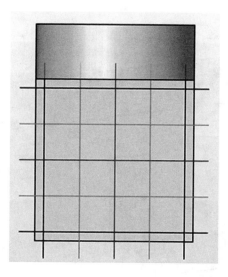

141. Create a sketch on the selected surface as shown in Figure 133.

Figure 133

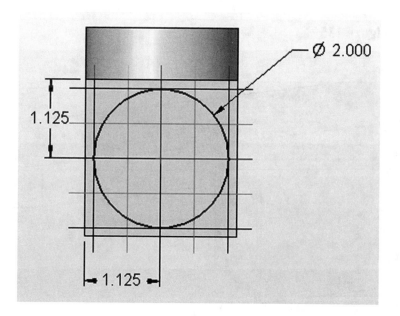

142. Exit out of the Sketch command and change the view to dimetric as shown in Figure 134.

Figure 134

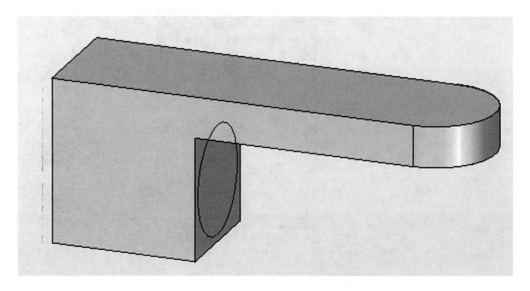

143. Use the **Cutout** command to extrude or "cut" out the circle that was just completed. Your screen should look similar to Figure 135.

Figure 135

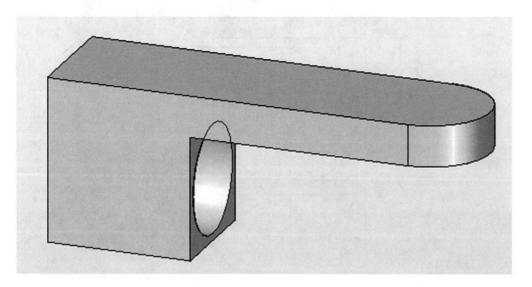

144. Move the cursor to the upper left portion of the screen and left click on **Sketch** as shown in Figure 136.

Figure 136

Left Click Here

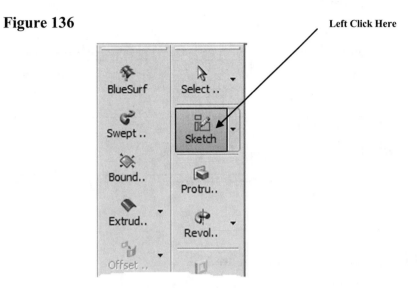

145. Left click on the top portion of the part causing it to turn red as shown in Figure 137.

Figure 137

Left Click Here

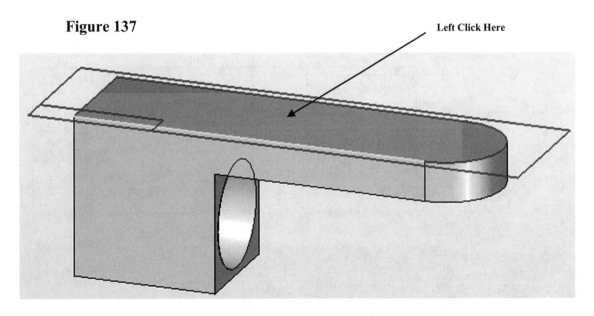

146. A perpendicular view will be displayed as shown in Figure 138.

Figure 138

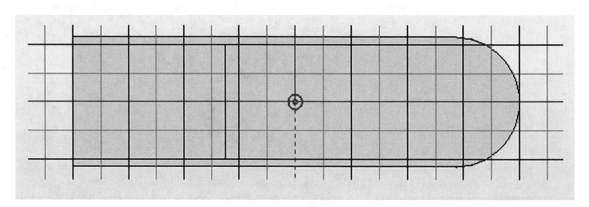

147. Draw the sketch shown in Figure 139.

Figure 139

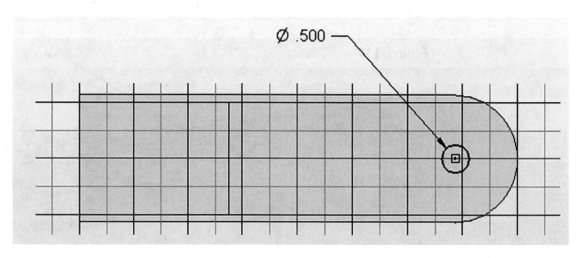

$\varnothing$ .500

148. Use the **Cutout** command to extrude or "cut" out the circle that was just completed. Your screen should look similar to Figure 140.

Figure 140

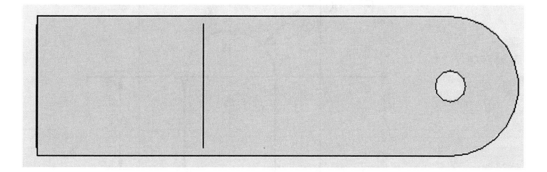

149. Change the view to dimetric. Your screen should look similar to Figure 141.

Figure 141

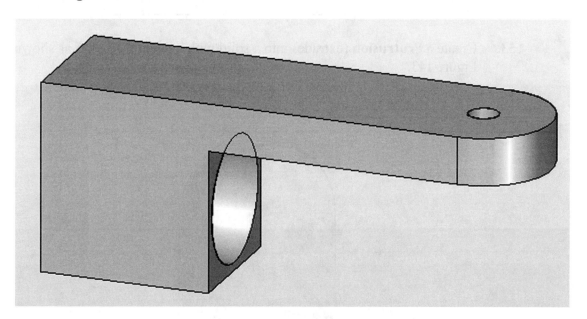

150. Save the part as **Pistoncase1.par** where it can be easily retrieved later.

151. Begin a new drawing.

152. Complete the sketch shown in Figure 142.

Figure 142

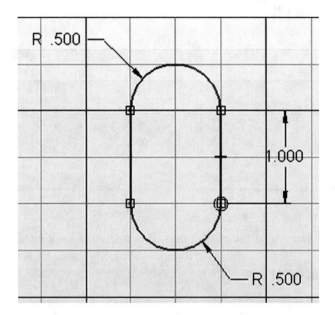

153. Create a **Protrusion** (extrude) into a solid with a thickness of **.25** as shown in Figure 143.

Figure 143

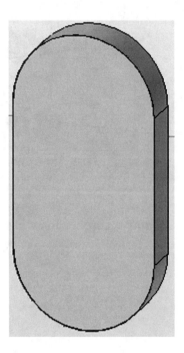

154. Complete the following sketch. Use the center of the outside radius as the center of the circle as shown in Figure 144.

Figure 144

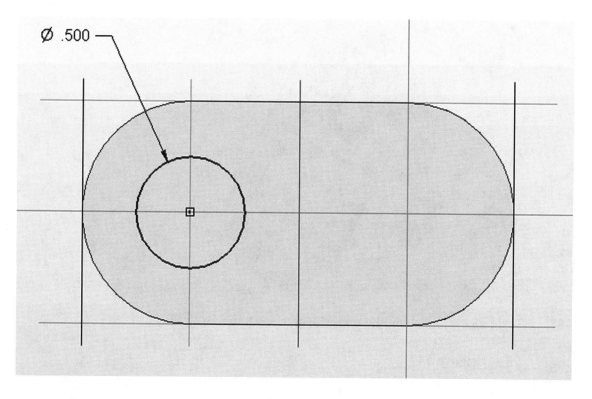

Ø .500

155. Create a **Protrusion** (extrude) into a solid with a thickness of **.25** as shown in Figure 145.

Figure 145

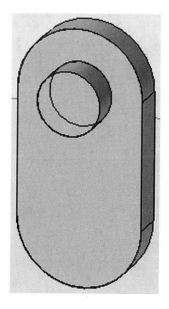

156. Use the rotate command to roll the part around to gain access to the opposite side as shown in Figure 146.

Figure 146

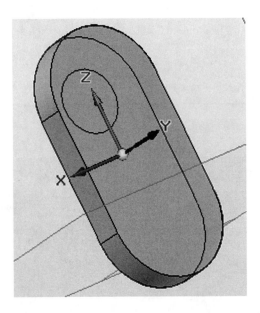

157. Begin a new sketch on the opposite side as shown in Figure 147.

Figure 147

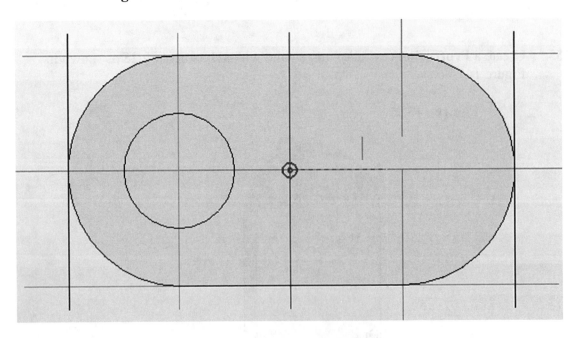

158. Complete the following sketch as shown in Figure 148.

      Figure 148

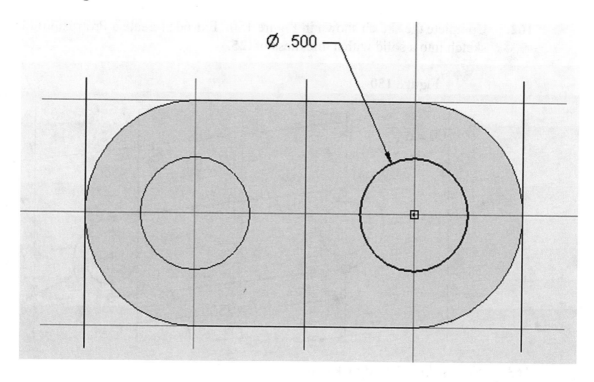

Ø .500

159. Extrude the sketch into a solid with a thickness of **.25** as shown in Figure 149.

      Figure 149

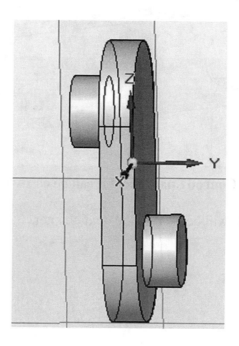

160. Save the part as **Crankshaft1.par** where it can be easily retrieved later.

161. Begin a new drawing.

162. Complete the sketch shown in Figure 150. Extrude (create a Protrusion) the sketch into a solid with a thickness of **.25**.

Figure 150

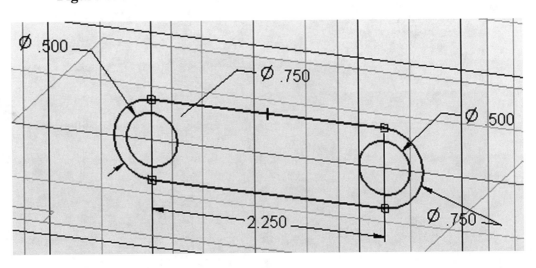

163. Your screen should look similar to Figure 151.

Figure 151

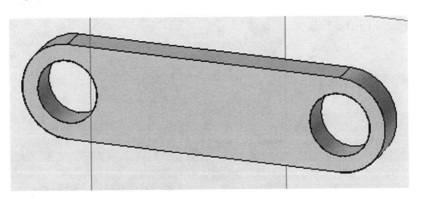

164. Save the part as **Conrod1.par** where it can be easily retrieved later.

165. All of these parts will be used in the next chapter.

Chapter 7 Introduction to Assembly View Procedures

Objectives:

- Learn to import existing solid models into an assembly
- Learn to assemble parts in an assembly
- Learn to edit/modify parts while in an assembly
- Learn to create a motor to simulate motion in an assembly

Chapter 7 includes instruction on how to construct the assembly shown below.

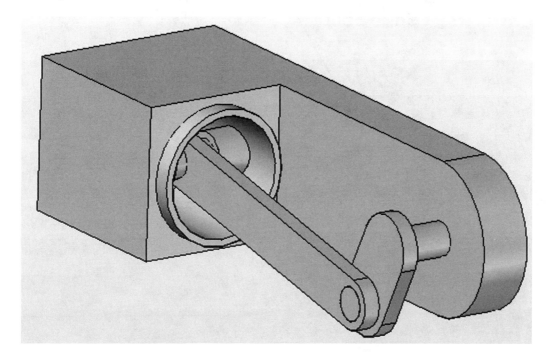

1.   Start Solid Edge by referring to "Chapter 1 Getting Started".

2.   After Solid Edge is running, begin an Assembly drawing. Move the cursor to the middle left portion of the screen and left on **Assembly** as shown in Figure 1.

Figure 1

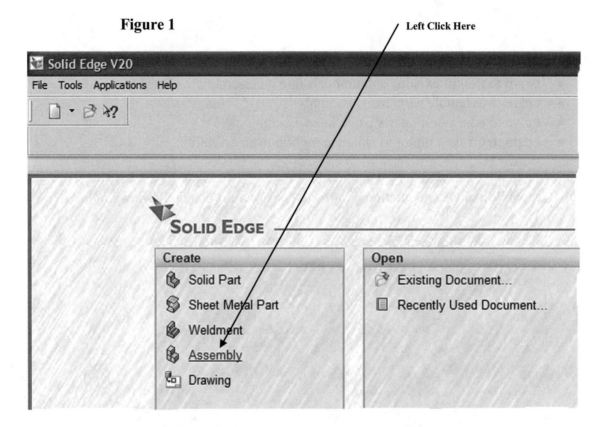

3. The Assembly portion of Solid Edge will open. Your screen should look similar to Figure 2.

Figure 2

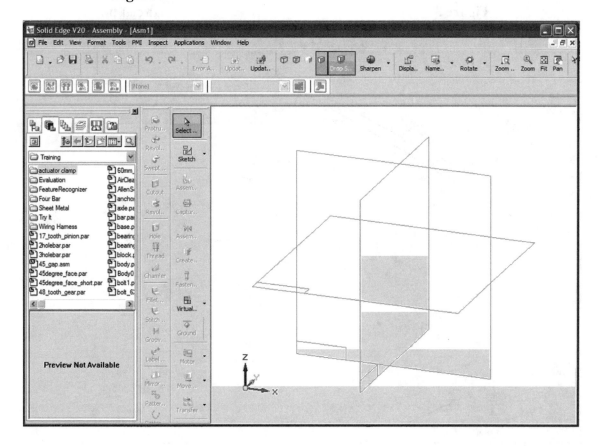

4. Move the cursor to the upper left portion of the screen and locate the parts that were created in Chapter 6. A preview of each of the highlighted parts will be displayed as shown in Figure 3.

Figure 3

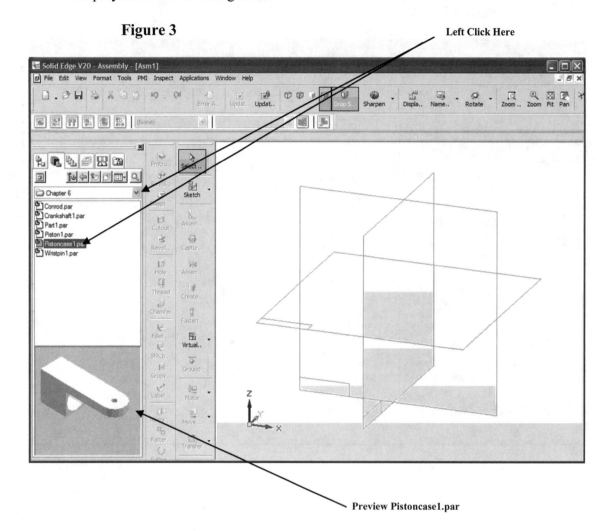

Left Click Here

Preview Pistoncase1.par

5.   Double click on **Pistoncase1**. Solid Edge will insert the part into the workspace
as shown in Figure 4.

Figure 4

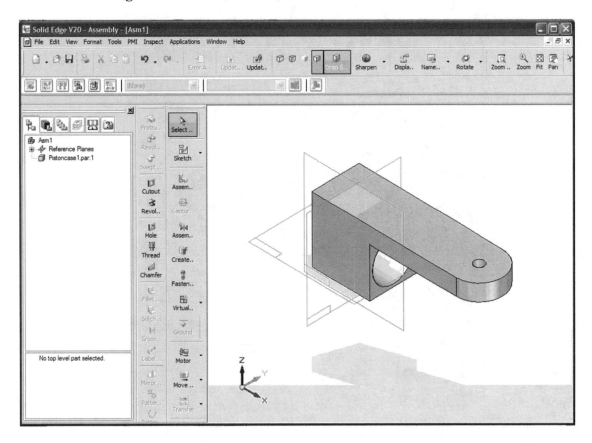

6.   Move the cursor to the upper left portion of the screen and left click on the "Parts Library" tab. Double click on **Wristpin1** as shown in Figure 5.

Figure 5

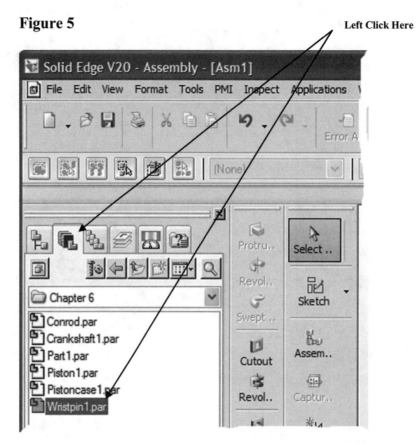

7.   Solid Edge will insert the part into the workspace as shown in Figure 6. To move the wristpin left click on it (holding the left mouse button down) and move it to the desired location.

Figure 6

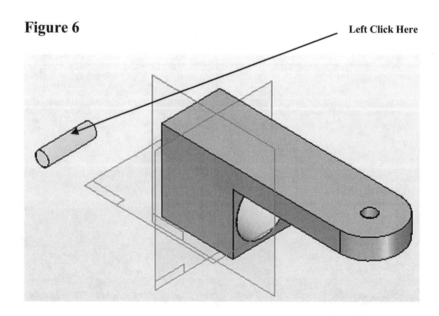

8. Move the cursor to the upper left portion of the screen and left click on the "Parts Library" tab. Double click on **Piston1** as shown in Figure 7.

Figure 7

Left Click Here

9. Solid Edge will insert the part into the workspace as shown in Figure 8. To move the piston left click on it (holding the left mouse button down) and move it to the desired location.

Figure 8

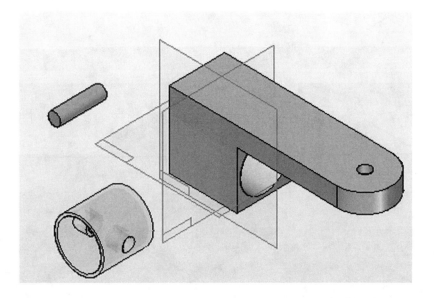

10.   Move the cursor to the upper left portion of the screen and left click on the "Parts Library" tab. Double click on **Conrod** as shown in Figure 9.

Figure 9   Left Click Here

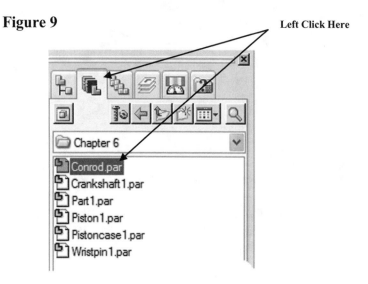

11.   Solid Edge will insert the part into the workspace as shown in Figure 10. To move the connecting rod, left click on it (holding the left mouse button down) and move it to the desired location.

Figure 10

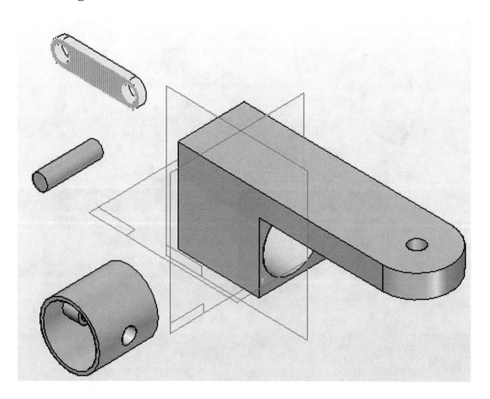

12. Move the cursor to the upper left portion of the screen and left click on the "Parts Library" tab. Double click on **Crankshaft1** as shown in Figure 11.

Figure 11

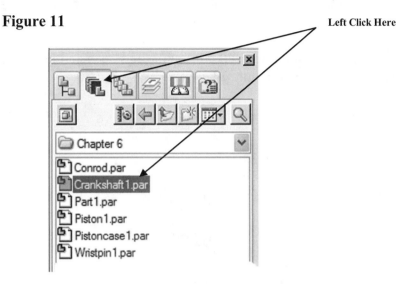

13. Solid Edge will insert the part into the workspace as shown in Figure 12. To move the crankshaft, left click on it (holding the left mouse button down) and move it to the desired location

Figure 12

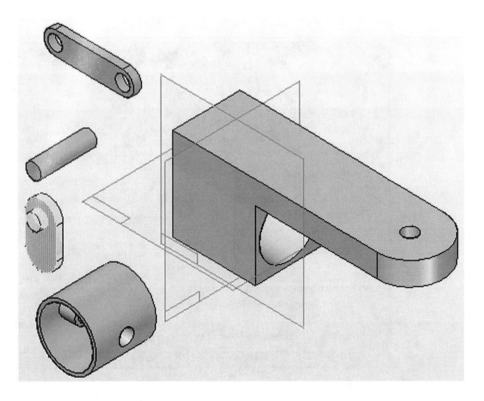

14. The first part inserted into the assembly becomes "grounded" meaning it cannot be moved without un-grounding it. All other parts can be moved using the Move Part command.

15. Move the cursor to the lower left portion of the screen and left click on **Move Part** as shown in Figure 13.

Figure 13   Left Click Here

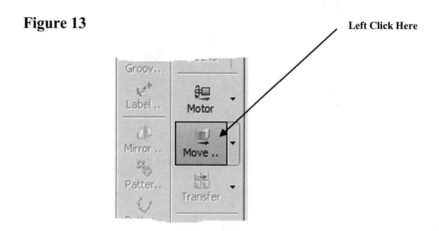

16. The Analysis Options dialog box will appear. Left click on **OK** as shown in Figure 14.

Figure 14   Left Click Here

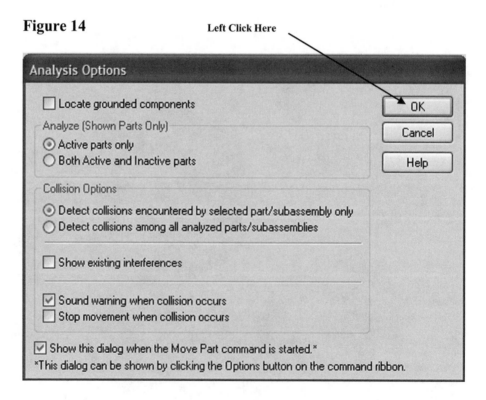

17. Move the cursor of the connecting rod causing it to turn red. Left click (holding the left mouse button down) and drag the part to the desired location as shown in Figure 15. The Move icon will have to be reselected before each part can be moved.

Figure 15　　　　　　　　Left Click Here Holding the Left Mouse Button Down

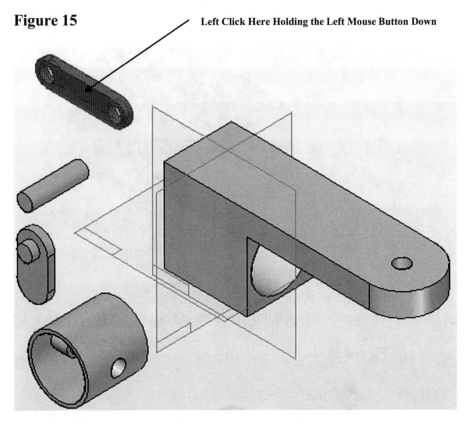

18. Move the cursor to the upper left portion of the screen and left click on **Assemble** as shown in Figure 16.

Figure 16　　　　　　　　　　Left Click Here

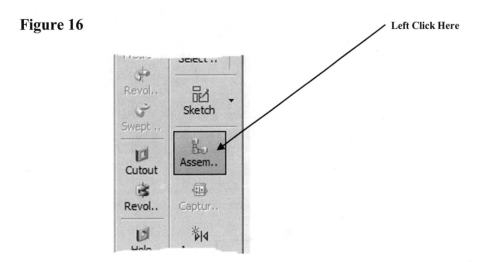

19. Move the cursor to the piston causing the OUTSIDE edges of the part to turn red. Left click once as shown in Figure 17. The piston will turn yellow.

Figure 17

Left Click Here

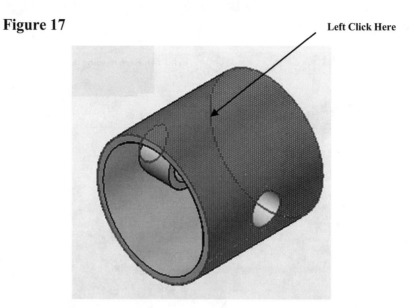

20. Move the cursor to the piston case causing the cylinder edges to turn red. Left click once as shown in Figure 18. The cylinder edges will turn yellow.

Figure 18

Left Click Here

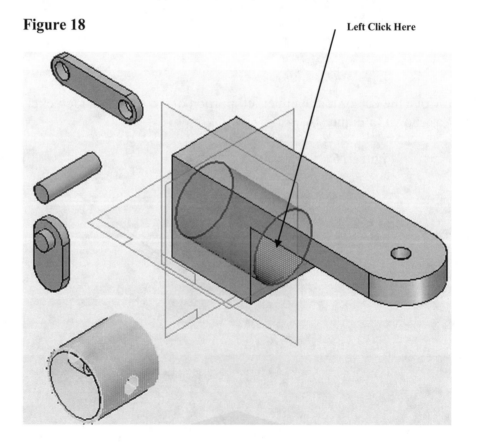

21. If the piston is facing the wrong way (top portion of the piston is facing downward) move the cursor to the upper middle portion of the screen and left click on **Flip** as shown in Figure 19.

Figure 19

Left Click Here

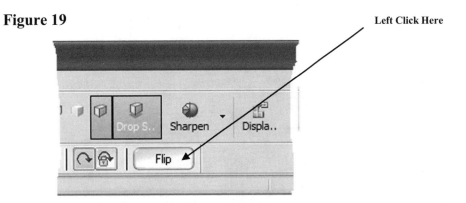

22. The open end of the piston should be facing downward as shown in Figure 20.

Figure 20

Open Portion of Piston

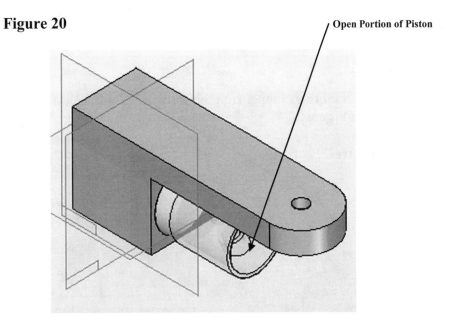

23. Right click anywhere around the part. Your screen should look similar to Figure 21.

Figure 21

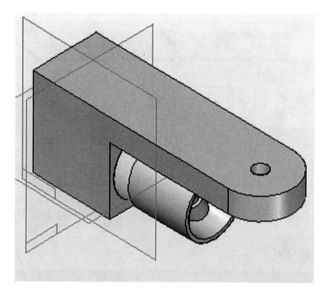

24. Move the cursor to the upper right portion of the screen and left click on **Rotate** as shown in Figure 22.

Figure 22

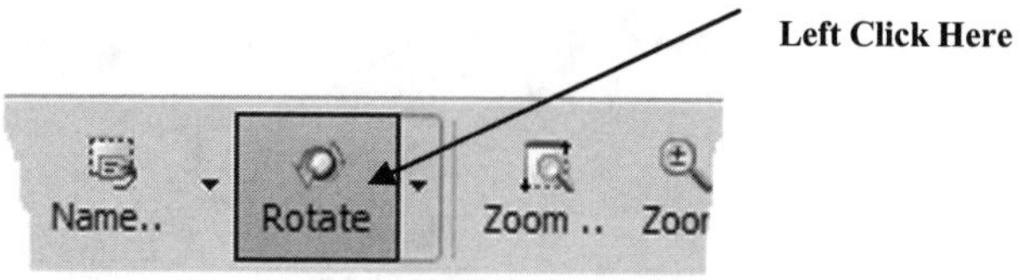

Left Click Here

25. Rotate the entire assembly upward as shown in Figure 23.

Figure 23

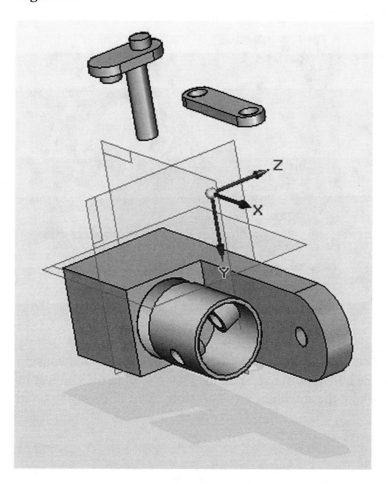

26. Move the cursor to the upper middle portion of the screen and left click on **Assemble** as shown in Figure 24.

Figure 24

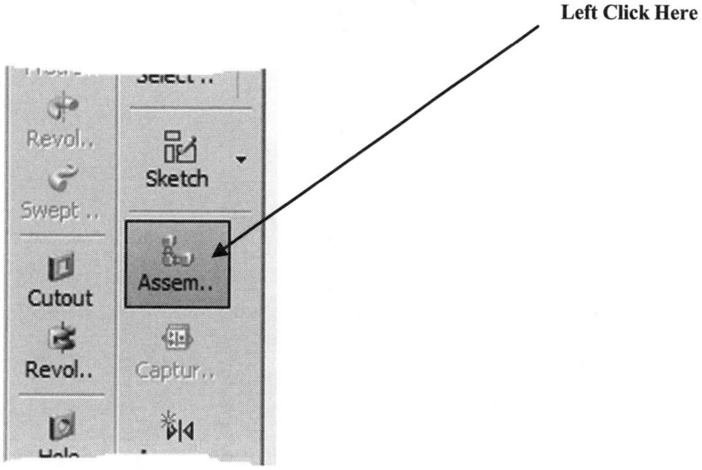

Left Click Here

27.   Move the cursor to the wristpin hole on the piston causing the inside edge of the hole to turn red. Left click once as shown in Figure 25.

Figure 25

Left Click Here

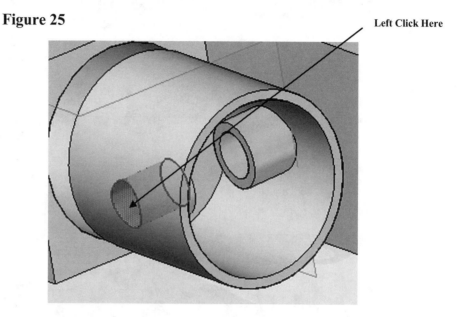

28.   Move the cursor to the upper portion of the connecting rod causing the inside edges of the hole to turn red. You may have to drag the connecting rod to a position where the end holes are accessible. Left click as shown in Figure 26.

Figure 26

Left Click Here

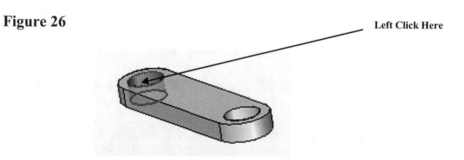

29. The connecting rod hole and the center of the wristpin hole in the piston will be aligned as shown in Figure 27.

Figure 27

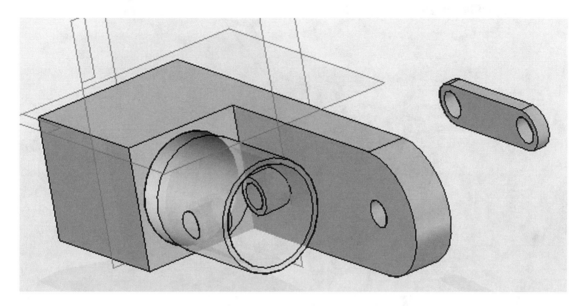

30. Right click anywhere around the part. Your screen should look similar to Figure 28.

Figure 28

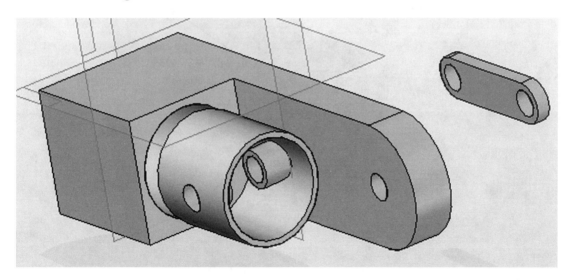

31. Use the Rotate command to rotate the entire assembly around to gain access to the underside of the piston. Use the Move command to move the connecting rod to the location shown in Figure 29. This will take some skill.

Figure 29   Move Connecting Rod Here

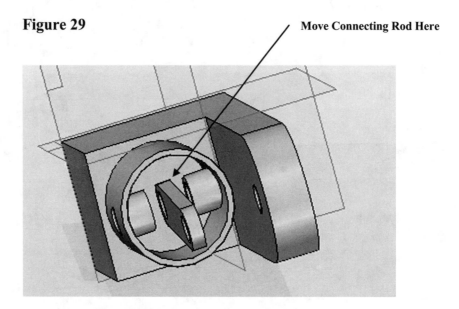

32. Move the cursor to the upper middle portion of the screen and left click on **Assemble** as shown in Figure 30.

Figure 30   Left Click Here

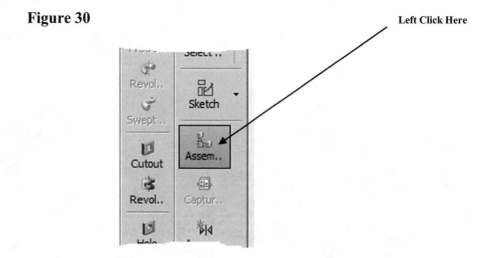

33. Move the cursor to the left side of the connecting rod causing the edges of the connecting rod to turn red. Left click as shown in Figure 31. You may have to zoom in so that Solid Edge will find the proper surface.

Figure 31

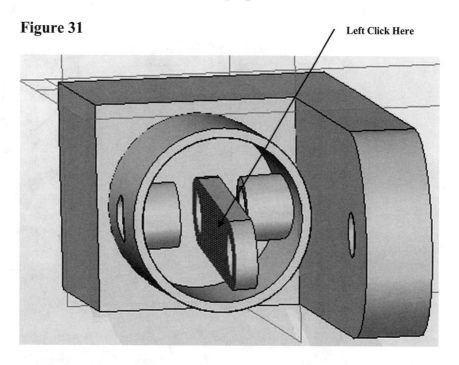

34. Move the cursor to the inside hole of the piston (while the surface is red) and left click as shown in as shown in Figure 32.

Figure 32

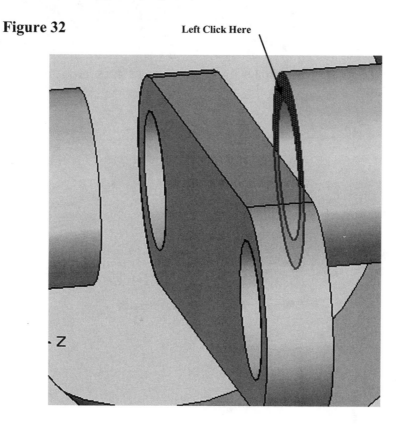

35. Solid Edge will move the connecting rod flush with the piston hole as shown in Figure 33.

Figure 33

Surfaces Flush

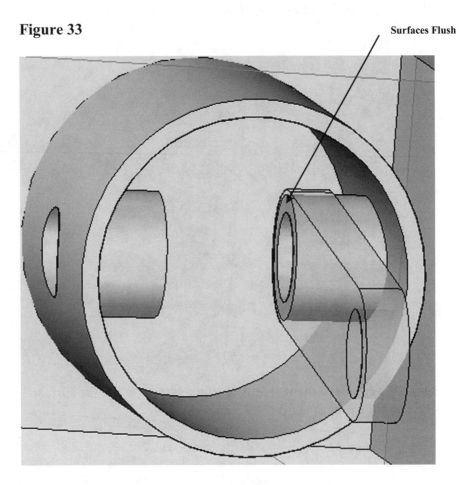

36. Move the cursor to the upper middle portion of the screen and enter **-.500** in the box shown in Figure 34.

Figure 34

Enter -.500 Here

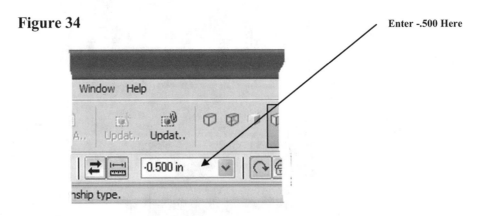

37. The connecting rod should be centered in the piston. Right click once. Your screen should look similar to Figure 35.

Figure 35

38. Move the cursor to the upper middle portion of the screen and left click on **Assemble** as shown in Figure 36.

Figure 36

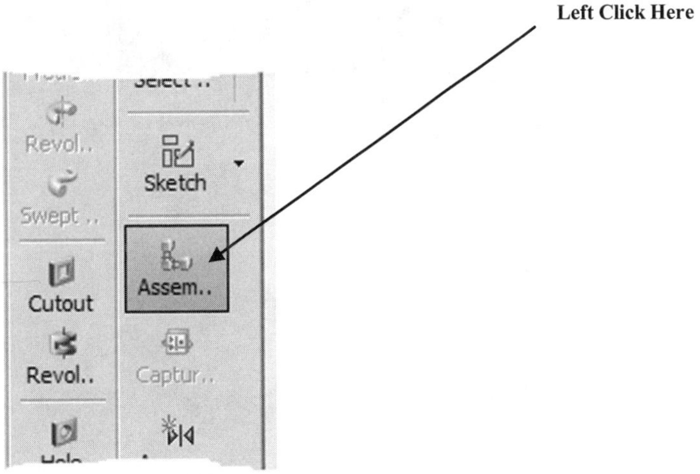

39. Move the cursor to the wristpin causing the edges of the wristpin to turn red. Left click once as shown in Figure 37.

Figure 37

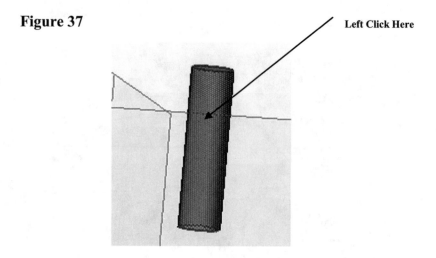

Left Click Here

40. Move the cursor to the wristpin hole causing the edges of the hole to turn red. Left click once as shown in Figure 38.

Figure 38

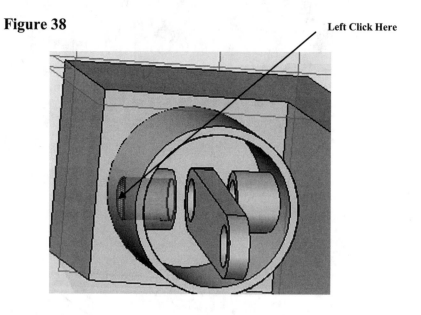

Left Click Here

41. The wristpin will be placed in the wristpin hole. Right click once. Your screen should look similar to Figure 39.

Figure 39

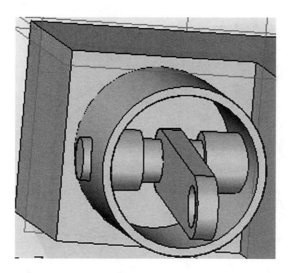

42. Move the cursor to the upper middle portion of the screen and left click on **Assemble** as shown in Figure 40.

Figure 40

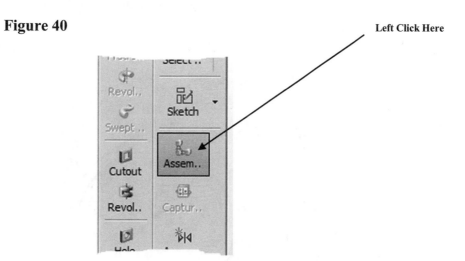

Left Click Here

43. Move the cursor to the side of the wristpin causing the surface of the wristpin to turn red. Left click once as shown in Figure 41.

Figure 41

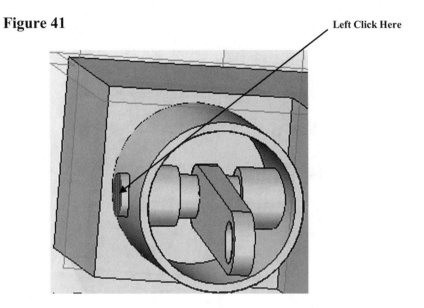

Left Click Here

44. Move the cursor to the side of the connecting rod causing the surface of the connecting rod to turn red. Left click once as shown in Figure 42.

Figure 42

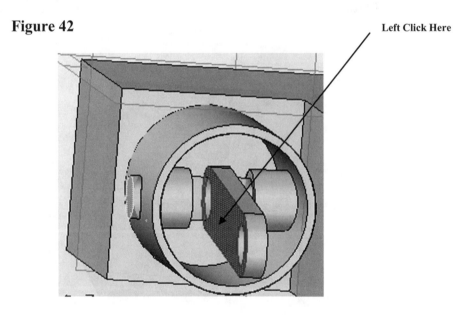

Left Click Here

45. Solid Edge will move the wristpin flush with the side of the connecting rod as shown in Figure 43.

Figure 43

Flush With Connecting Rod Edge

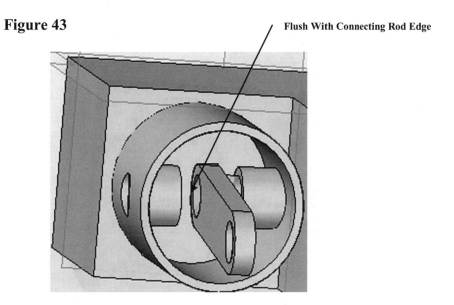

46. Move the cursor to the upper middle portion of the screen and enter **-.813** in the box shown in Figure 44.

Figure 44

Enter -.813 Here

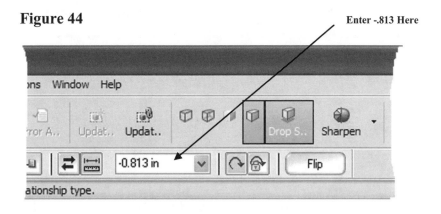

47.   Solid Edge will center the wristpin in the wristpin hole. Right click once. Your screen should look similar to Figure 45.

Figure 45

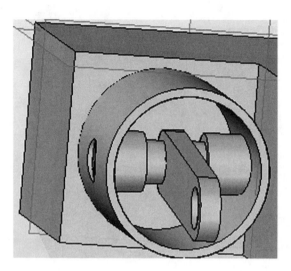

48. Use the zoom command to zoom out making the crankshaft visible as show in Figure 46.

Figure 46

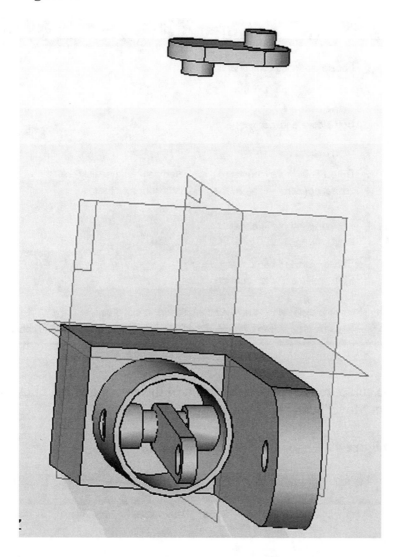

49. Move the cursor to the lower middle portion of the screen and left click on **Move Part** as shown in Figure 47.

Figure 47   Left Click Here

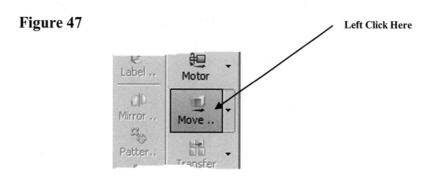

375

50. The Analysis Options dialog box will appear. Left click on **OK** as shown in Figure 48.

Figure 48

Left Click Here

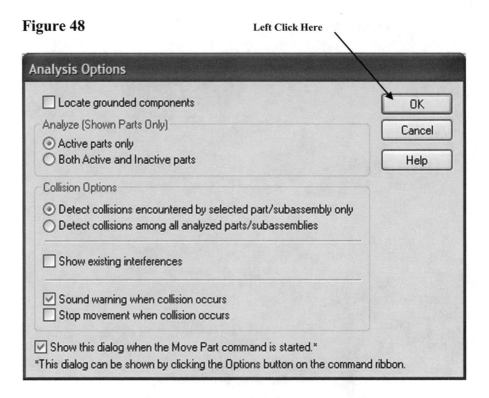

51. Move the cursor to the upper left portion of the screen and left click on the "Freeform Move" icon as shown in Figure 49.

Figure 49

Left Click Here

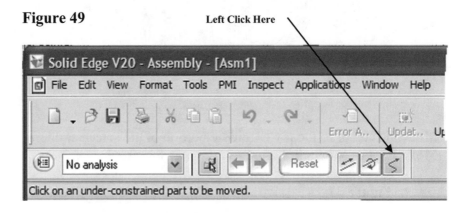

52. Move the cursor over the crankshaft causing it to turn red. Left click (holding the left mouse button down) as shown in Figure 50.

Figure 50

Left Click Here

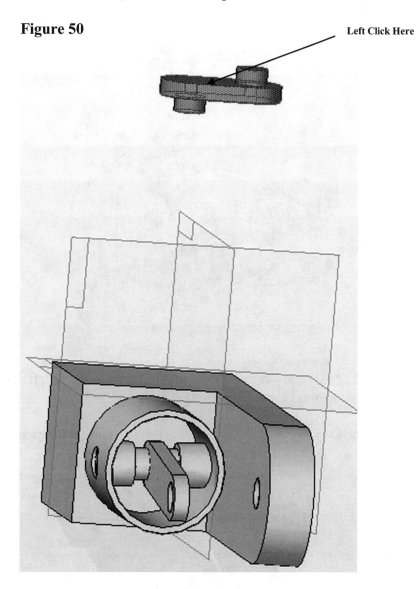

53. Move the crankshaft closer to the piston assembly as shown in Figure 51.

Figure 51

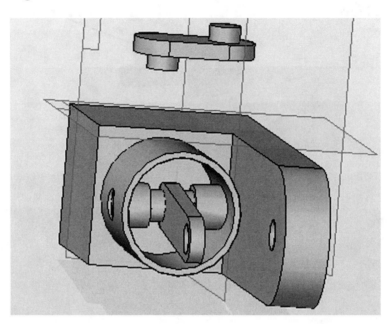

54. Move the cursor to the lower middle portion of the screen and left click on **Assemble** as shown in Figure 52.

Figure 52

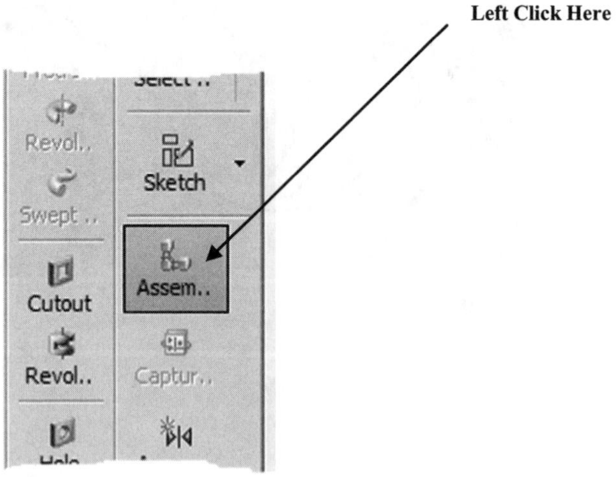

55. Move the cursor to the crankshaft pin that will secure the connecting rod. The edges of the crankshaft pin will turn red. Left click once as shown in Figure 53.

Figure 53

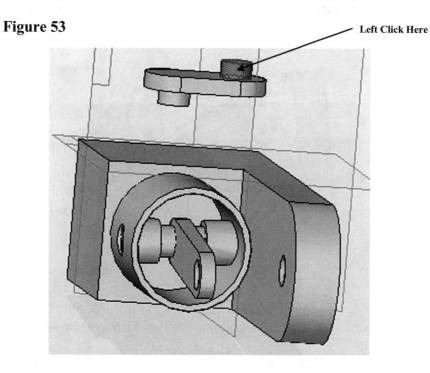

Left Click Here

56. Move the cursor to the connecting rod hole that will be secured to the crankshaft. The edges of the connecting rod hole will turn red. Left click inside the hole as shown in Figure 54.

Figure 54

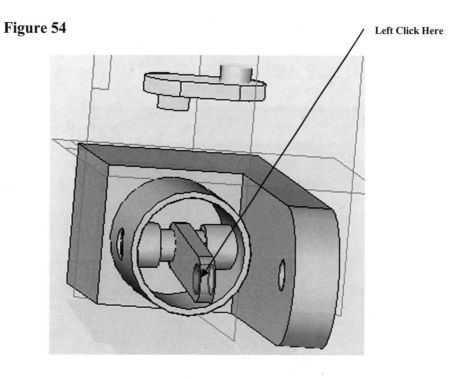

Left Click Here

57. Solid Edge may insert the crankshaft backwards as shown in Figure 55.

Figure 55   Crankshaft is Backwards

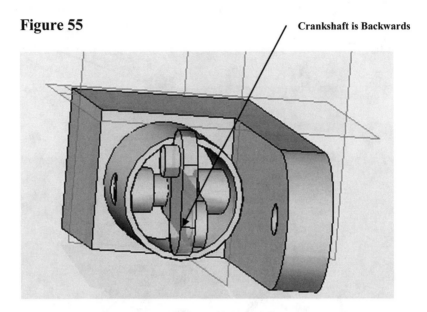

58. If Solid Edge inserted the crankshaft backwards move the cursor to the upper middle portion of the screen and left click on **Flip** as shown in Figure 56.

Figure 56   Left Click Here

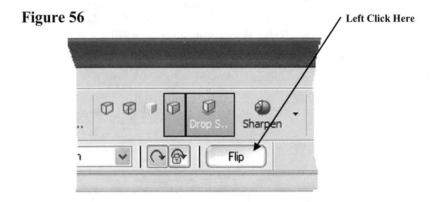

59. Solid Edge will flip the crankshaft around as shown in Figure 57.

Figure 57   **Crankshaft Flipped Over**

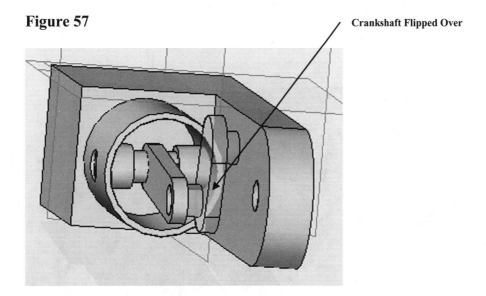

60. Right click anywhere around the part. Your screen should look similar to Figure 58.

Figure 58

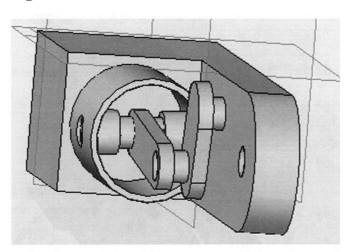

61. The center of the crankshaft pin and the center of the connecting rod are now aligned.

62.   Use the rotate command to rotate the assembly to better access the side of the crankshaft as shown in Figure 59.

Figure 59

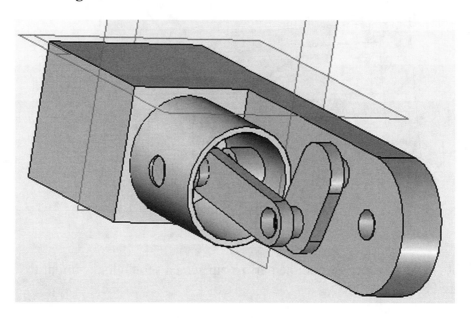

63.   Use the move command to move the crankshaft closer to the connecting rod as shown in Figure 60.

Figure 60   Move Parts Together

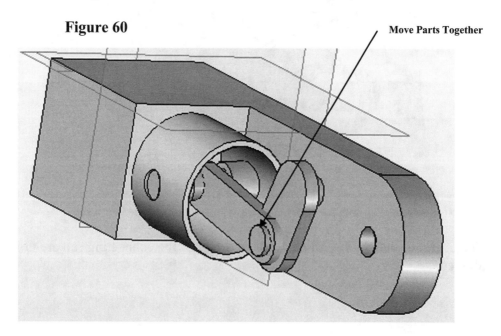

64. Move the cursor to the upper middle portion of the screen and left click on **Assemble** as shown in Figure 61.

Figure 61

Left Click Here

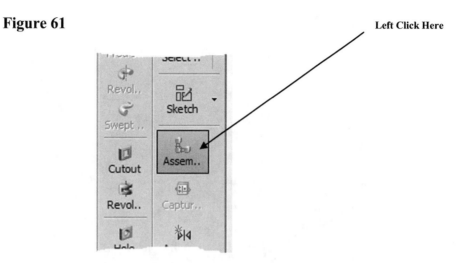

65. Move the cursor to the crankshaft pin causing the edge of the pin to turn red. Left click as shown in Figure 62.

Figure 62

Left Click Here

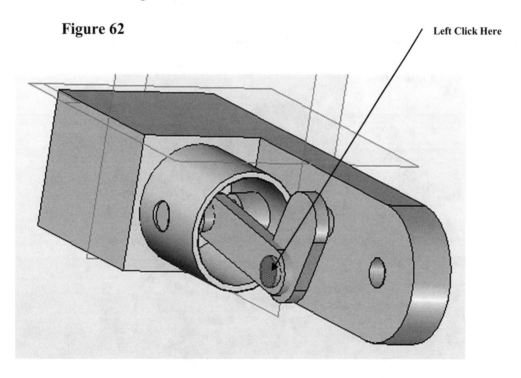

66.   Move the cursor to the connecting rod side. The edge of the connecting rod will turn red. Left click then right click as shown in Figure 63.

Figure 63   Left Click/Right Click Here

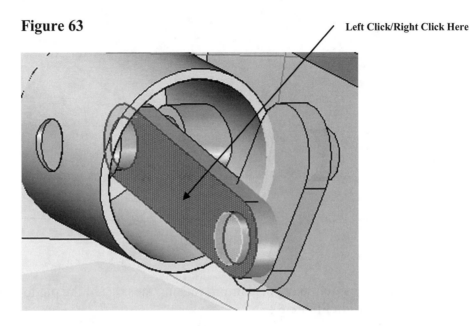

67.   Your screen should look similar to Figure 64.

Figure 64

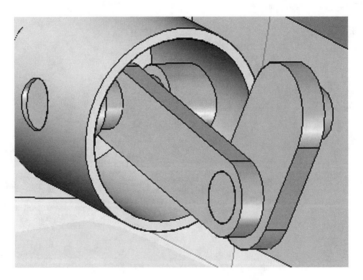

68. Move the cursor to the upper middle portion of the screen and left click on **Assemble** as shown in Figure 65.

Figure 65

Left Click Here

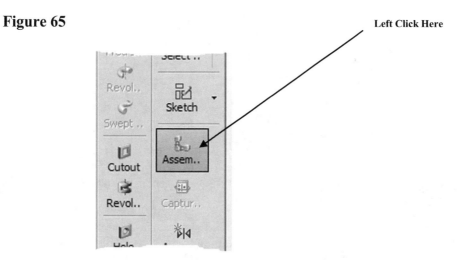

69. Move the cursor to the crankshaft pin that will be secured in the piston case. The edges of the pin will turn red. Left click once as shown in Figure 66.

Figure 66

Left Click Here

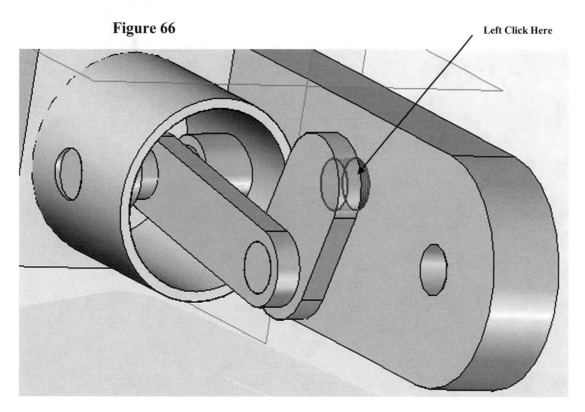

70.   Move the cursor to the piston case hole that will secure the crankshaft. The edges of the hole will turn red. Left click then right click as shown in Figure 67.

Figure 67   Left Click/Right Click Here

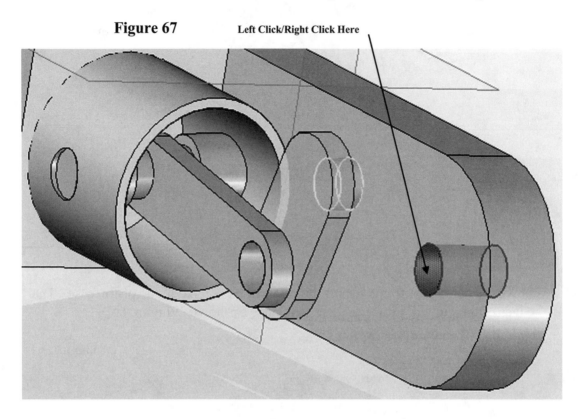

71. Solid Edge will place the crankshaft pin into the piston case as shown in Figure 68.

Figure 68

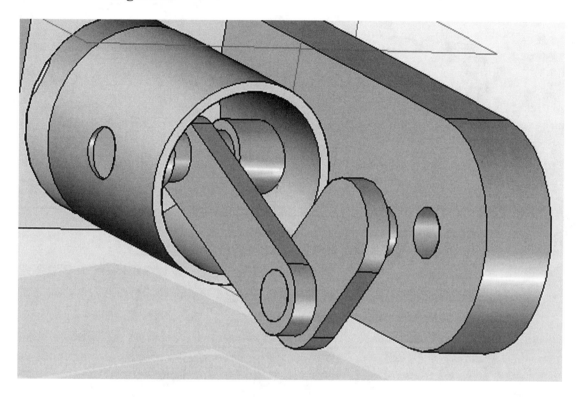

72.   Use the Rotate command to roll the parts around to view the assembly. Your screen should look similar to Figure 69.

Figure 69

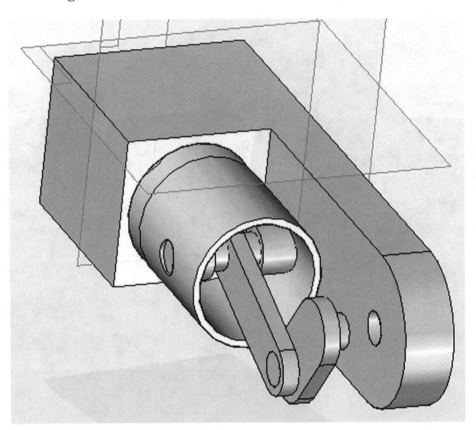

73. The length of the connecting rod must be modified. Move the cursor over the connecting rod causing it to turn red as shown in Figure 70. Left click causing it to turn yellow.

Figure 70

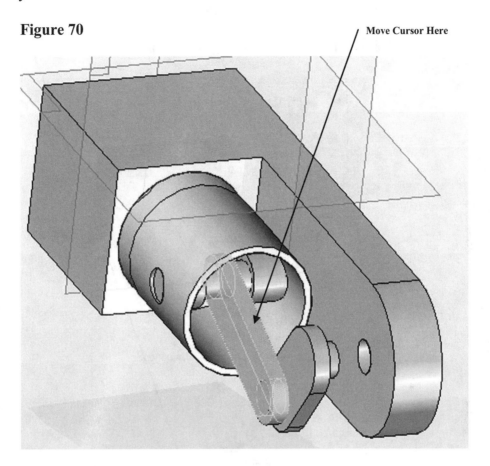

74.   Move the cursor to the upper left portion of the screen and left click on the Assembly PathFinder tab. Right click on **Conrod.par:1**. A pop up menu will appear. Left click on **Edit** as shown in Figure 71.

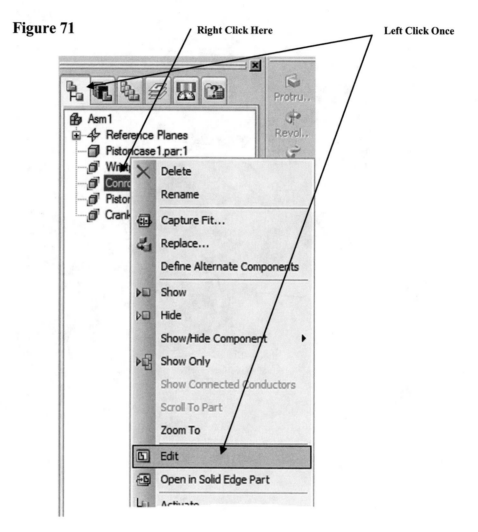

Figure 71

75.   Solid Edge will return to the sketch area as shown in Figure 72.

Figure 72

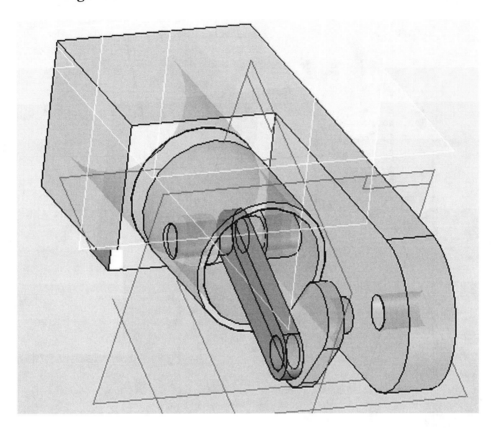

76.   Move the cursor to the upper left portion of the screen and right click on **Sketch 1**. A pop up menu will appear. Left click on **Edit Profile** as shown in Figure 73.

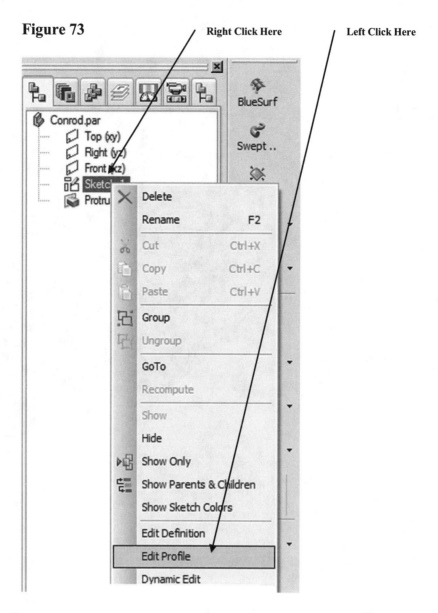

Figure 73

77. The sketch used to create the connecting rod will be displayed. Solid Edge will rotate the sketch around. Move the cursor over the 2.25 dimension and single left click as shown in Figure 74.

Figure 74

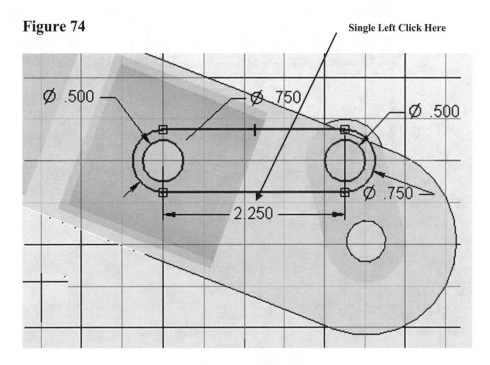

78. Enter **4.5** in the dimension box as shown in Figure 75. Press the **Enter** key on the keyboard.

Figure 75

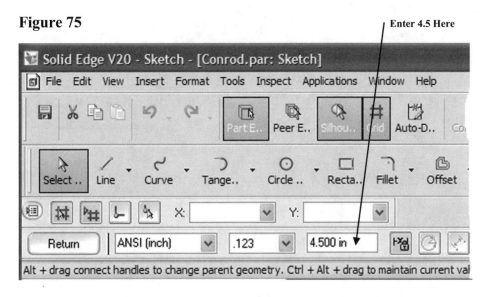

79.   Move the cursor to the upper left portion of the screen and left click on **Finish** as shown in Figure 76.

Figure 76

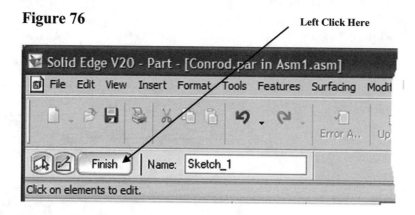

80.   Your screen should look similar to Figure 77.

Figure 77

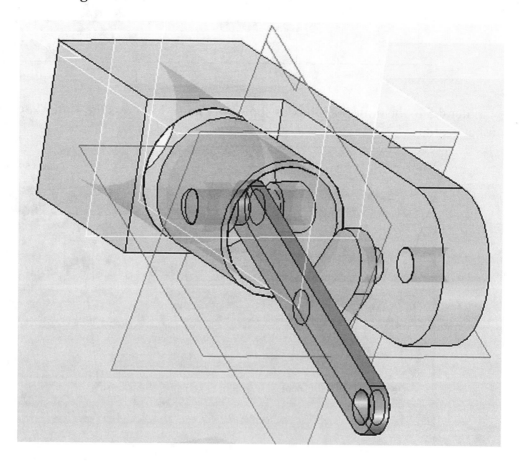

81. Move the cursor to the upper left portion of the screen and left click on **File**. A drop down menu will appear. Left click on **Close and Return** as shown in Figure 78.

Figure 78

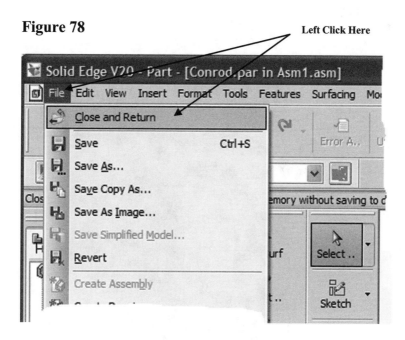

82. The length of the connecting rod will become 4.5 inches as shown in Figure 79.

Figure 79

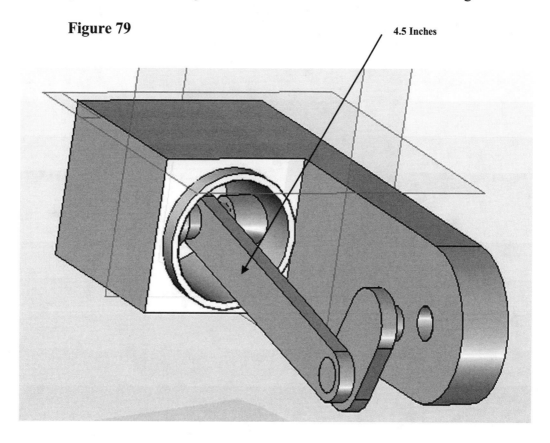

83. The length of the crankshaft pin also must be modified. Use the Rotate command to rotate the assembly to gain access to the backside of the crankshaft as shown in Figure 80.

Figure 80

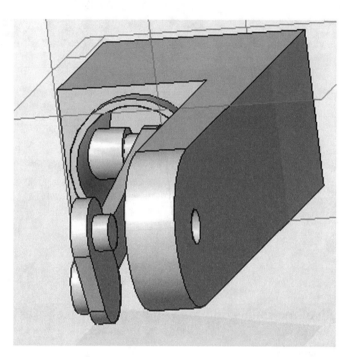

84. Move the cursor to the upper left portion of the screen and right click on **Crankshaft1.par:1**. A pop up menu will appear. Left click on **Edit** as shown in Figure 81.

Figure 81

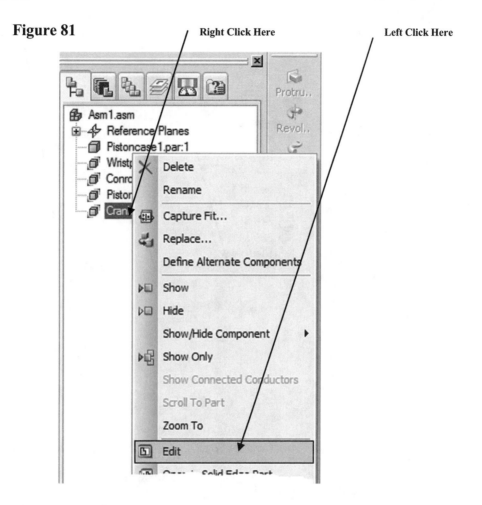

85. Move the cursor to the upper left portion of the screen and right click on **Protrusion 3**. A pop up menu will appear. Left click on **Edit Definition** as shown in Figure 82.

Figure 82

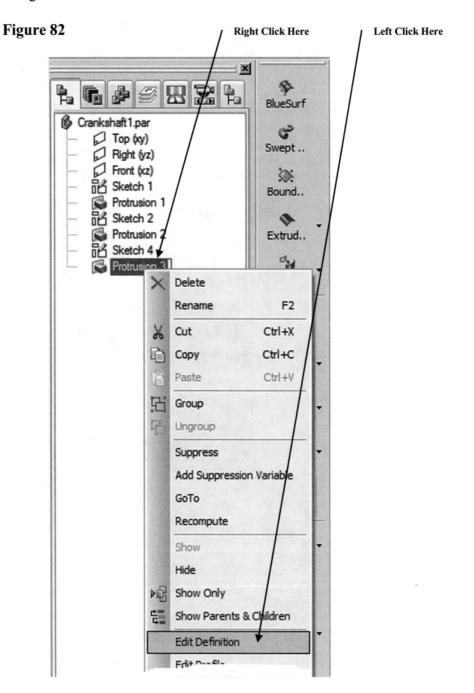

86. Move the cursor over the .250 dimension and left click once as shown in Figure 83.

Figure 83

Left Click Here

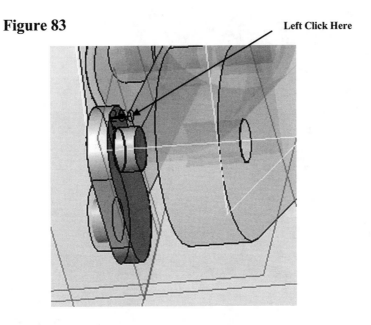

87. Move the cursor to the upper middle portion of the screen and enter **1.750** in the dimension box as shown in Figure 84. Press the **Enter** key on the keyboard.

Figure 84

Enter 1.750 Here

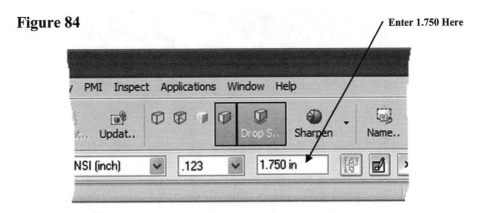

88.   A preview of the protrusion will be displayed as shown in Figure 85.

Figure 85

Preview of 1.75 Protrusion

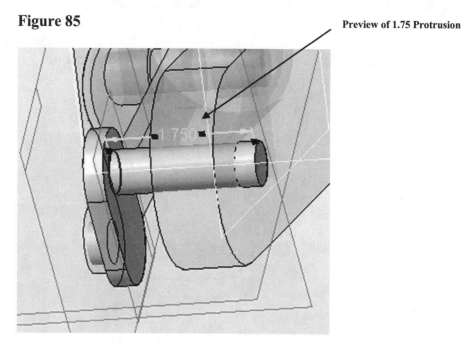

89.   Move the cursor to the upper middle portion of the screen and left click on **Finish** as shown in Figure 86.

Figure 86

Left Click Here

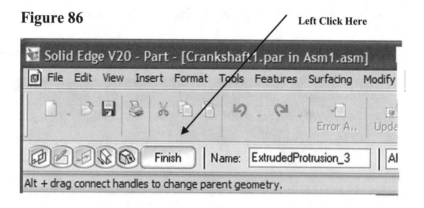

90.   Move the cursor to the upper middle portion of the screen and left click on **File**. A drop down menu will appear. Left click on **Close and Return** as shown in Figure 87.

Figure 87

Left Click Here

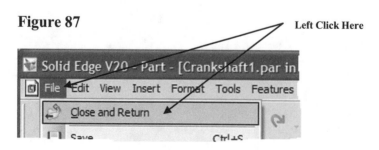

91. The changes made to the crankshaft will be displayed. Your screen should look similar to Figure 88.

Figure 88

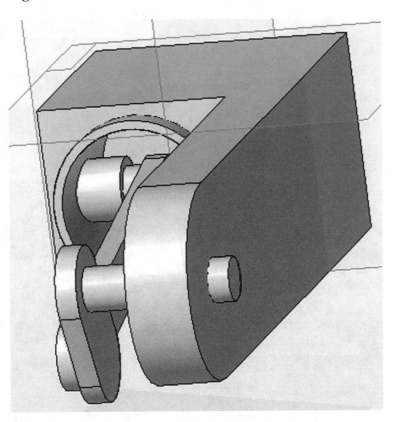

92. Use the Rotate command to position the part as shown in Figure 89.

Figure 89

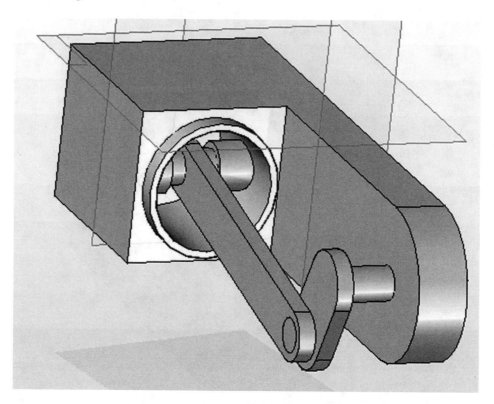

93. Hide the top, front and side planes as previously described in Chapter six.

94. Move the cursor to the lower right portion of the screen and left click on the drop down arrow next to the "Motor" icon. A drop down box will appear. Left click on **Motor** as shown in Figure 90.

Figure 90   **Left Click Here**

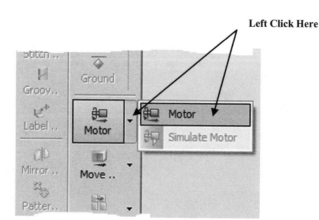

95. Move the cursor to the crankshaft pin that protrudes through the piston case. The crankshaft pin will turn red as shown in Figure 91.

Figure 91

Turned Red

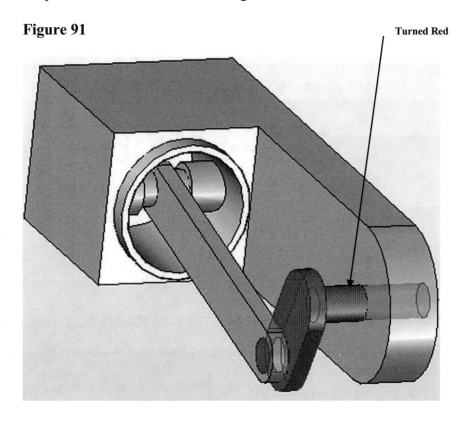

96. Left click <u>twice</u>. A blue arrow will be displayed showing the direction of motion as shown in Figure 92.

Figure 92

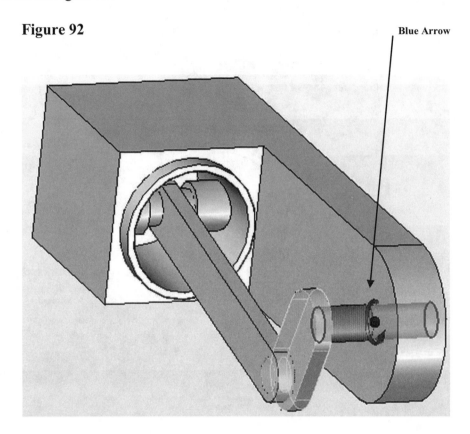

97. Move the cursor to the upper left portion of the screen and left click on **Finish** as shown in Figure 93.

Figure 93

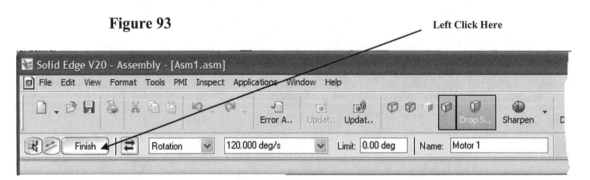

98. Move the cursor to the lower left portion of the screen and left click on the drop down arrow next to the "Motor" icon. A drop down menu will appear. Left click on **Simulate Motor** as shown in Figure 94.

Figure 94

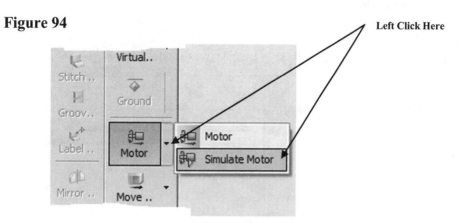

99. The Motor Group Properties dialog box will appear. Left click on **OK** as shown in Figure 95.

Figure 95

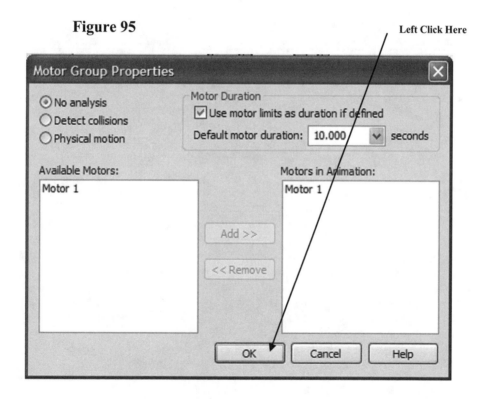

100. Move the cursor to lower middle portion of the screen and left click on the "Play" button as shown in Figure 96.

Figure 96

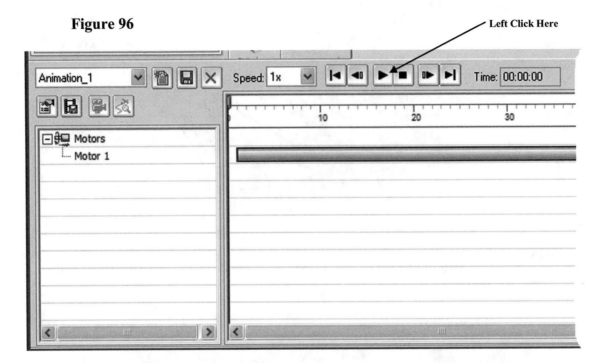

406

101. Solid Edge will begin animating the part. Use the "Rewind" button to repeat the simulation. Your screen should look similar to Figure 97.

Figure 97

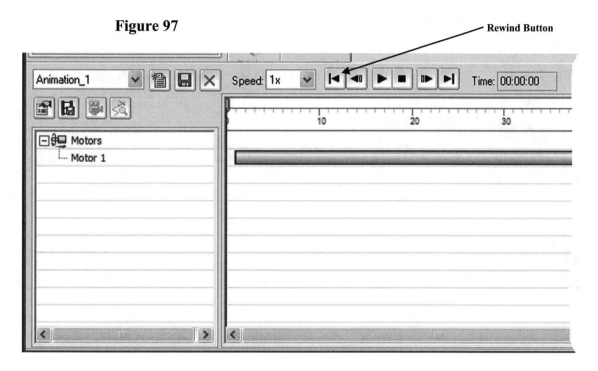

102. To delete the displayed animation move the cursor to the lower left corner of the screen and left click on the red **X** as shown in Figure 98.

Figure 98

Left Click Here

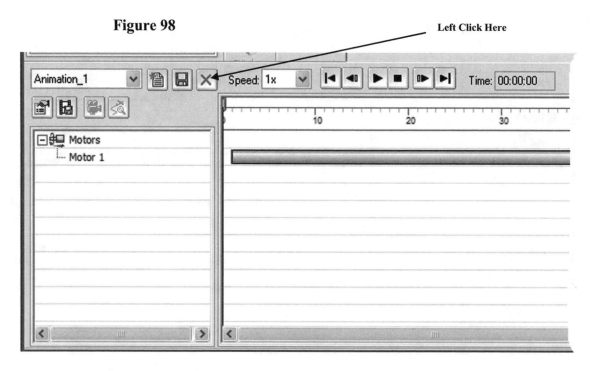

103. The Delete Animation dialog box will appear. Left click on **Yes** as shown in Figure 99.

Figure 99

Left Click Here

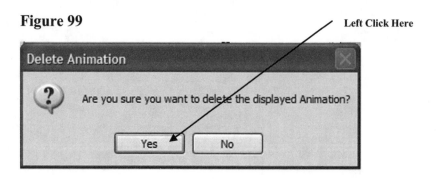

104. Move the cursor to the upper left portion of the screen and left click on **Select** as shown in Figure 100. The Display Animation dialog box will close.

Figure 100

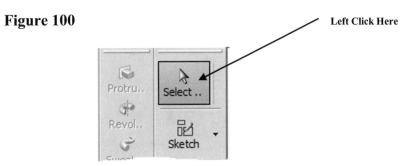

Left Click Here

105. Save the part where it can easily be retrieved. When closing Solid Edge, a dialog box will appear indicating that models making up the assembly have been modified. Left click on **Yes** as shown in Figure 101.

Figure 101

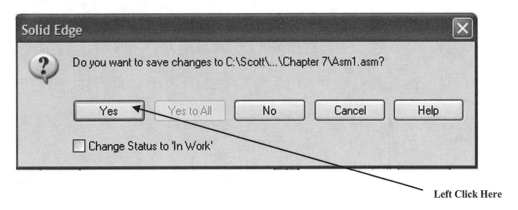

Left Click Here

Index